Praise for
The Social Lives of Birds

"A tour-de-force survey of how birds live their lives—with all the drama, surprise, humour, sadness and amazement of any human soap-opera"

—Stephen Moss, author of *Ten Birds That Changed the World*

"Utterly fascinating. A book which opens a door into the world of birds and reveals an astonishing array of behaviour in the ways birds feed, nest and flock. Strassmann is the perfect guide to this world: an author as much fascinated by the science and research as she is motivated by the sheer joy and wonder of the birds themselves"

—James Macdonald Lockhart, author of *Raptor*

"The main features of birds most of us are interested in concern their feathers, flight, nesting, feeding, foraging, mating, predator evasion, migration, and group vs. solitary behavior. I know of no other book that so thoroughly covers the hugely extensive scientific literature from the experts who spend their lives and fortunes on their work"

—Bernd Heinrich, author of *Mind of the Raven*

"In this elegant and masterful treatment of avian life, Strassmann makes it abundantly clear that the proverb 'birds of a feather flock together' is one massive understatement. Birds variously pair up, lek, roost, form colonies, team up to assist the parent, breed communally, and turn super-social. She will intrigue the novice while transporting even the most knowledgeable bird lover in fresh and unexpected directions"

—Mark Moffett, author of *The Human Swarm*

"Birds of a feather not only flock together, but sleep, feed, migrate, mate, and raise young together, too. Sometimes birds move about and live together with only their own species and sometimes they are in mixed flocks. Strassmann, a world-leading scientist on the communal lives of diverse lineages of life on Earth, clearly explains the benefits and costs of the different ways in which birds spend time together. Her easy-to-follow writing is based on scientific findings from peer-reviewed literature, and it takes us from parasitic cowbirds in the Americas to penguins in Antarctica and drongos in India. The world, as she explains, is a more interesting place, because we humans share so much with birds when it comes to living and loving together"

—Mark Hauber, author of *Bird Day*

"Strassmann has distilled the enormous scientific literature on bird social behavior and made it available to all of us. Cooperate or compete? Live alone or join a colony? Favor relatives or non-relatives? Choose immediate gains or invest in the future? Shaped by natural and kin selection, and the environments they live in, birds navigate all these challenges in extraordinarily diverse ways"

—Ellen Ketterson, author of *Snowbird*

"By weaving together her own personal stories with the ecological and evolutionary insights gleaned from long-term studies of avian species from around the world, Strassmann introduces us to the wonderfully complex and varied social lives of birds. This book is a must read for anyone interested in learning about not only why the birds we often see or hear behave the way that they do, but also how the scientists who have studied them for the past century helped uncover these social secrets"

—Dustin Rubenstein, author of *Animal Behavior*, 12th Edition

"In vivid explanations of complex bird behavior, Strassmann makes the ordinary—vultures roosting in a tree, swallows soaring by a riverbank—extraordinary. She also makes the extraordinary—prancing prairie chickens, albatross living in colonies of thousands—understandable, using the lens of evolution. From American Robins to Taiwan Yuhinas, this book is a celebration of birds and the scientists who study them, both working harder than we can imagine"

—Marlene Zuk, author of *Paleofantasy*

"A leading expert on social wasps and social amoebae brings to bear her expertise and passion to the social world of birds. The result is a fascinating guided tour of the many mysteries of bird life and glimpses of their probable solutions—why birds flock during the day, sleep in communal roosts at night, forage with other species, nest in crowded colonies, breed communally, mate in leks, or outside the pair bond, care for young not their own, and quarrel with their siblings and parents"

—Raghavendra Gadagkar, author of *Survival Strategies*

"Why do birds join flocks, nest in sprawling colonies, or tend to eggs that are not their own? Using global examples, from mixed-species groups in Finland to families of Pinyon Jays in the American southwest, Strassmann carefully unpacks the motivations behind avian social behaviors. A mix of personal observation, interviews with experts, and a review of the scientific literature, this is a fascinating and informative read"

—Jonathan C. Slaght, author of *Owls of the Eastern Ice*

"Strassmann effortlessly weaves the communal lives of birds with the science and the scientists whose studies have revealed the drivers of their behavior. The stories she tells provide fascinating insights into the world of animal behavior"

—Walter Koenig, coauthor of *Ecology and Evolution of Cooperative Breeding in Birds*

The Social Lives of Birds

Flocks, Communes, and Families

Joan E Strassmann

Illustrations by Anthony Bartley

HEADLINE PRESS

First published in the US in 2025 by Tarcher
An imprint of Penguin Random House LLC

First published in the UK in 2025 by Headline Press
An imprint of Headline Publishing Group Limited

1

Cataloguing in Publication Data is available from the British Library

Hardback ISBN 978 1 0354 1713 1
Trade Paperback ISBN 978 1 0354 1714 8

Book design by Shannon Nicole Plunkett

Offset in 12.4/17.3pt Adobe Caslon Pro by Six Red Marbles UK, Thetford, Norfolk

Printed and bound in Great Britain by Clays Ltd, Elcograf S.p.A.

MIX
Paper | Supporting responsible forestry
FSC® C104740

Headline's policy is to use papers that are natural, renewable and recyclable products and made from wood grown in well-managed forests and other controlled sources. The logging and manufacturing processes are expected to conform to the environmental regulations of the country of origin.

Headline Publishing Group Limited
An Hachette UK Company
Carmelite House
50 Victoria Embankment
London EC4Y 0DZ

The authorised representative in the EEA is Hachette Ireland, 8 Castlecourt Centre, Dublin 15, D15 XTP3, Ireland (email: info@hbgi.ie)

www.headline.co.uk
www.hachette.co.uk

For my parents,
Wolfgang Paul Strassmann and
Elizabeth Marsh Fanck Strassmann,
who showed me the natural world and
encouraged me to explore

CONTENTS

Introduction:
Are Birds Social? — 1

Chapter 1
Flocks: Can Many Do Better Than Few? — 7

Chapter 2
Communal Roosts: Why Sleep Together? — 25

Chapter 3
Mixed-Species Flocks:
Follow the Alarm Caller — 45

Chapter 4
Colonies: Safe Places near Food — 69

Chapter 5
Seabird Colonies: How to Rear Young
by the Largest Larder on Earth — 97

Chapter 6
Leks: Where Males Dance and Females Choose — 115

Chapter 7
Mate Choice and Parental Care:
Competition in the Family — 139

Chapter 8

Families with Helpers: Older Siblings,
Lonely Bachelors, and More — 169

Chapter 9

Communal Nesters: Confusion in the Nest — 191

Chapter 10

Supersocial Groups: Birds That Are Always Together — 209

Conclusion:
Why Are Birds So Social? — 247

Acknowledgments — 251

Notes — 255

Index — 283

The Social Lives of Birds

Introduction

Are Birds Social?

My morning walk to our local park began with the pounding of a Northern Flicker on a neighbor's metal chimney, calling its mate and advertising its territory. Then I saw a flock of Common Grackles leave one tree and fly to another. On this warm March morning, the American Robins were already paired up and flying together (the male identifiable by his jet-black head). The Dark-eyed Juncos breed much farther north, but here they were, practicing their high-pitched trills, flashing their white tail feathers, and foraging in flocks.

Bird-watching has a rather staid reputation, in which the birder is thought to be a quiet, bookish observer, carefully noting species and habitats with the quasi-academic bent of an ardent plane-spotter. But unlike planes, birds are social, and their lives are full of drama, intrigue, cooperation, and competition. They skirt danger, form couples and communities, sneak around and steal, and flirt like their lives depend on it. All this makes bird-watching fascinating—more like picking up a juicy beach read or tuning in to a thrilling miniseries than recording the airplane tail numbers at a local runway.

This book celebrates the social lives of birds, from the southernmost penguins, petrels, and albatrosses, through the tropical fairywrens and drongos, and up to the Arctic gannets, puffins, and guillemots during the breeding season. Of course, stories of birds would be incomplete without stories of the researchers and observers who have uncovered their secrets; their adventures (and my own) round out this look at birds around the globe.

Social living takes a lot of different forms in birds. Each chapter takes on a different type of sociality. I begin with flocks, those assemblages of foraging birds in which each searches for food and tries not to become prey. Birds in flocks follow rules that have evolved to keep them safe. They balance staying safe in groups with finding more food away from groups. However, in groups they can share vigilance, which gives birds more time for foraging and can stymie predators.

Like all animals, birds sleep. This brings me to communal roosts. When birds sleep in communal roosts, there is safety in numbers, providing them with better protection from predators. Some birds, like Black Vultures, also learn at roosts and follow others to a ripe carcass. Small birds, like Long-tailed Tits, snuggle together to stay warm as they roost.

Sometimes birds of different species flock together, with each species taking a different role. These groups can be so stable that their territorial boundaries are preserved across decades, even as individual members are slowly replaced. Birds often congregate around active forager species. Listening for their alarms helps other birds, who might have their beaks deep in a bark crevice, realize danger has arrived. Mixed-species foraging flocks are found all over the world, so if you look for them, you will see them.

Groups of nesting birds are called colonies. For example, Cliff Swallows nest in groups under bridges and culverts and have been studied for decades. Breeding in a colony means there is more of a chance a neighbor will steal nesting material, sneak an egg into your nest, or even mate with your partner, but there are considerable advantages, again generally around avoiding predators.

Seabird colonies can have thousands of individuals. They have the special challenge of eating far from where they nest because the oceans are the world's largest larder. Atlantic Puffins form colonies of thousands in the far north. Penguin colonies are iconic. Many isolated islands in the tropics could be called bird islands, home to frigate birds, gulls, and others.

Leks are arenas where groups of males show off and females choose a mate. Lekking species have no paternal care, for the males shed any responsibility for the young, instead competing with one another in some of the most flamboyant displays in the animal world, from the eyespots of the peacocks to the dances of the Greater Sage-grouse. Females often choose the same "best" male. But it is not all honest display on the lekking ground—Ruff males may mimic females and sneak in a mating right on the dancing male's territory. But neither is it all competition as male Lance-tailed Manakins have dancing wingmen that never themselves achieve a mating.

Even birds that do not feed in flocks or nest in colonies have a social stage, for all bird babies hatch from eggs and then nearly always need parental care. Pairing up and parental care are the most basic forms of sociality in birds. The energy requirements of birds mean that most bird families have two parents that tend the young. But these partners do not show the kind of devoted monogamy you might imagine. While birds pair with their mates, and in some species the males follow their mates around attentively in mating season, additional breeding outside the pair bond is more the rule than the exception. Nor, despite the ties of kinship, is there complete harmony between parents and offspring, or among the offspring themselves. These intrigues can make for great stories.

In harsh conditions, two parents are not always enough to rear the young, so they recruit young from earlier years to help with the brood. Sometimes stray males also chip in, perhaps hoping to inherit breeding rights should the mated father perish. Australia is home to the most species of families with helpers, though perhaps the best known are Florida Scrub-jays.

There is a surprising kind of group nesting that might seem to make little sense, since parents cannot keep track of their own young. It is communal breeding and is found in Greater Anis. Several pairs build a nest together and lay eggs in it. One male takes the risky role of incubating the eggs at night, and all feed the young and sound the alarm when a predator nears. But it is not all cooperation and brotherhood with these birds, for females toss some of each other's eggs out of the shared nest.

Finally, some birds—the supersocial groups—have it all, never forgoing a chance to be with others. White-winged Choughs of Australia come to mind, as do Pinyon Jays of the American West. These birds feed in flocks, nest in colonies, and care for the young at least partly communally. They are wicked smart, often deceptive, and have intrigued researchers for decades.

There are four themes that run through these chapters and help us understand avian sociality. First, birds act in ways that increase their own ability to survive and reproduce. Second, when they act in ways that look cooperative, it is either because they themselves get a tangible return or because they contribute to the next generations by helping relatives with their young. This is generally called kin selection and explains a lot of apparently altruistic behavior. Third, different environments favor different kinds of behaviors, and how variable those environments are is also important. Fourth, evolution acts on what has gone before, with traits always building on the traits of the ancestors. Taken together, these four themes give us the colonial penguins in Antarctica, communal yuhinas in Taiwan, helpers in Long-tailed Tits in Europe, supersocial Pinyon Jays in the American West—all the diversity that is the avian world.

Our planet is immeasurably richer because of its birds, but this richness is under threat.[1] The better we understand the needs of birds, the more likely we will be to conserve habitats on which birds depend. I hope this book will help readers understand the actions and the needs of social birds. I hope when readers see a social bird not covered in this book, with a little observation they will be able to classify the kinds of interactions they see. And I hope the book helps everyone enjoy the birds.

CHAPTER 1

Flocks

Can Many Do Better Than Few?

Since flocks are the most common form of avian sociality spotted on a morning walk, they seem an appropriate place to start this book. A flock is any group of two or more birds that stays together for more than a moment, typically while they forage for food or fly to a new place.

Flocks are everywhere. Cedar Waxwings cluster in a hackberry tree whistling their high song. Brown Pelicans beat slowly along the Gulf Coast in a shallow V. American Coots paddle through the water eating vegetation in apparently companionable groups. A flock of Red-winged Blackbirds flies low over my local park, headed for the fields west of town. It seems like flying solo is not for the birds.

In general, there are two possible advantages to being in a group: protection from predators and enhanced discovery of food. There are also disadvantages serious enough to make grouping troublesome: Flocking increases competition for food and facilitates disease and parasite transmission. So why have so many birds landed on group living?

One compelling paper—with the equally compelling title "Geometry for the Selfish Herd"—changed the way we saw foraging flocks. Bill Hamilton published it in the *Journal of Theoretical Biology* in 1971.[1] In it, he makes two main points.

First, individuals in a group behave for their own interests and not for the good of the overall group. Flocking, then, is selfish rather than altruistic behavior. It helps individual birds survive. For biologists, this first point is not so surprising, because only characteristics that are genetically selfish will be passed down to the next generation. A bird that joined a flock to benefit someone else would not have as many babies as one that kept its own interests front and center, although a side effect of group-joining could be that it helps others, too.

The second point is captured in Hamilton's clever use of the word *geometry*, in which Hamilton reminds us not just of high school math but also of location. For any individual there will be better and worse places to be, and those depend in large part on where others are.

Hamilton's idea is that flocks form as the result of individuals adjusting their position to distance themselves from actual or potential predators. Each individual benefits from putting someone else between them and the approach path of a predator. The sum result of their efforts leads naturally to clustering.

Clustering is only one way to avoid predators. Others include dilution, confusion, vigilance, and defense. *Dilution* means that a single bird's chances of being a predator's meal are lower if the predator has a lot of other birds to choose from, and *confusion* means that it might be harder for a predator to single out one bird from many. The last two ways group members might avoid predators are more active. *Vigilance* means that individuals in a group can take turns watching for predators while others forage. The more individuals in a group, the less time individual birds must spend raising their heads to look for predators instead of eating. Finally, *defense* means that a group of birds can attack a predator, mobbing it and chasing it away.

One of the best-studied flocking birds is the Common Redshank, *Tringa totanus*, commonly seen feeding on estuarine mudflats and salt marshes in Britain. Redshanks breed coastally throughout Asia and Europe.[2] They are sandpipers, sharing the genus *Tringa* with Willets, Wandering Tattlers, and Greater and Lesser Yellowlegs, among others. This genus is one of my favorites for shorebirds, though I love them all, especially because they are often waders, either on shore or land. They are elegant and striking with their bright long legs. The Common Redshank has the best legs of all, orange verging on red. I like its vertical stripes of brown on tan and its long bill, dark at the tip, orange at the base.

Redshanks stand out because they like to eat particularly small prey. At Tyninghame Estuary in Scotland, redshanks eat a lot of sandhoppers, *Orchestia*,[3] while in Teesmouth, on the coast of northeastern England, they favor a shrimplike amphipod, or mud scud, called *Corophium volutator*. Individually, these organisms, only about a fifth of an inch long, are not much of a meal, and so each redshank must eat many of them to survive. Other birds foraging in the area are smaller, like Dunlin, or larger, like Grey Plovers and Bar-tailed Godwits, but take larger prey.[4] More than other birds, redshanks must spend virtually every moment foraging to fill their bellies.

If food availability were all that mattered, however, one would expect redshanks to stick to the salt marshes, where their preferred prey are most abundant. But they do not. Redshanks prefer to forage in the less prey-rich mudflats and mussel beds, so much so that adults actively exclude juveniles from these hunting grounds. Why?

The answer forms part of a story written across many papers and kindly told to me by Will Cresswell, who began his academic career on the Tyninghame salt marshes and mudflats near the University of Edinburgh, where he was a student. Put simply, the redshanks are not the only birds looking for a meal. It turns out that the mudflats may have fewer of the amphipods they hunt, but they are farther from the forest cover used by Eurasian Sparrowhawks, which hunt Common Redshanks. Cresswell found that, compared with a redshank on the mudflats and mussel beds, a redshank on the salt marsh was nearly five times more likely to be killed by a sparrowhawk.[5]

Still, redshanks must hunt on the salt marshes, too—remember, they need an awful lot of sandhoppers and mud scuds to survive. How, then, do redshanks form flocks in ways that minimize their vulnerability to sparrowhawks while still getting enough to eat in the dangerous but productive salt marshes? Cresswell spent three years with binoculars and a telescope watching redshanks on the chilly Tyninghame Estuary. His goal was to find out if redshanks in larger flocks were less vulnerable to sparrowhawk attacks, and whether birds in larger flocks got more or less to eat.[6] Across eighty-nine censuses, Cresswell found that nearly all redshanks were in groups (of at least two dozen, but also as many as a hundred birds). The few birds that foraged alone joined groups at the first opportunity.

A sparrowhawk attack is always a surprise, for they launch without warning from cover where they are hard to see. They fly fast and hard toward their target, grabbing it with their talons before the prey is even aware of the pursuit. In Cresswell's research, sparrowhawks attacked lone birds and flocks of all sizes, but they were most successful catching lone birds and those in smaller flocks. This is a strong reason to be in a flock—the bigger, the better.

Clearly birds in larger flocks are safer, but within those flocks, what is the best position? According to Hamilton's "Geometry for the Selfish Herd," the best place to be is close to a neighbor. It reminds me of that old trope about how fast you have to run to escape a lion: faster than your neighbor.

Instead of finding some lions to chase them, John Quinn

and Cresswell, his advisor, headed back into the field to see where in the flocks Common Redshanks were safest.[7] They videotaped wooded areas near the salt marshes for six hundred hours before they had enough clear footage of sparrowhawk attacks on redshank flocks. A sparrowhawk's hunting strategy makes it ideal for this kind of study. Sparrowhawks attack from a hideaway, choose a single target, and don't change targets after launching. It was essential for this study to collect exact information not only on flock size but also on the exact positions of each bird and its neighbors at the time of an attack. In particular, the investigators wanted to see if sparrowhawks tended to target those redshanks that were farthest from their neighbors. Just as Hamilton's theory predicted, Quinn and Cresswell found that redshanks that wandered away from their neighbors were more likely to be attacked. Redshanks not only need to be in flocks, but they also need to stay close to their neighbors.

Another way redshanks reduce predation is by looking for sparrowhawks. This vigilance has a cost, since the redshank's head has to be up and looking instead of down and eating. It's easy to imagine that the more birds there are in the group, the less often any one bird would have to search for predators; just one or a few could be vigilant, communicating any danger (something that could be as simple as the redshank taking flight). Alex Sansom and Will Cresswell looked into this by banding and videotaping about seventy redshanks.[8] As the investigators predicted, the more birds in the flock, the less time any individual bird had to spend scanning for predators.

One downside to being in a larger flock is that these birds have to take more steps to get food. With other flock members probing the sand, and the sand shrimp fleeing this probing, the redshanks have to move. Indeed, birds in larger flocks spend more time foraging, time made available by the lessened need for vigilance.[9]

Common Redshanks face more problems than just predators. After all, they cannot nest in tidal flats, which are underwater at high tide. The meadows that are their favored nesting grounds are a bit more inland, in just the areas where farmers like to let their cattle graze. The large feet, incessant munching, and copious cowpies that come along with cattle are severe threats to a comparatively tiny redshank protecting its eggs. To document the effects, Lucy Malpas led a study of Common Redshank breeding populations on British salt marshes.[10] Her team found only 11,946 pairs on these marshes in 2011 where there had been 21,431 in 1985. The population had nearly halved—was it due to cattle?

Lack of nesting habitat is one of the main causes of this decline, making a bird that has "common" in its very name a species of conservation concern in the UK and Europe. Common Redshanks breed on grasslands and pastures, and nearly half of all the individuals in the species nest on salt marshes in Britain. They do best on lightly grazed sites, compared with ungrazed or heavily grazed ones. Eighty-four percent of the breeding pairs breed in East Anglia, mostly in an area called the Wash, a bay into the North Sea that defines the blunt thumb of East Anglia, about 100 miles (160 kilometers) north of London. Four rivers, the Welland, Witham, Nene, and Great Ouse, drain into the Wash.

Common Redshank declines were less severe in East Anglia than in other places because cattle grazing was lighter, but declines still happened. Elwyn Sharps led a team to figure out whether cattle were ever compatible with redshank conservation.[11] It is a complicated question, since light grazing by cattle, roughly one cow per 2.5 acres (1 hectare), actually helps the birds by keeping out woody species and generating the kind of patchy habitat favored by breeding redshanks. This means cattle must be managed at densities favorable for the birds and not the cattle owners, which will always create tension.

But there are other issues. The cattle in a given field do not necessarily distribute themselves evenly across the landscape. It turns out that the upland areas of salt marshes favored by breeding redshanks are the very areas cattle most favor in exactly the same season when the birds are nesting.

An effective solution would be to delay introduction of cattle to the upland salt marshes until mid-July, when the redshanks are done breeding. Alternatively, cattle owners could rotate cattle into and out of salt marshes, leaving some areas entirely cattle-free while the cattle graze in the others. The problem with this is that the highly site-faithful redshanks would still attempt to nest in the cattle-dense salt marshes.

Other solutions are unlikely to work. Sheep eat the vegetation too low. Horses trample more than cattle do. Researchers have concluded that to protect the birds, it is key to measure cattle density properly and to keep them off the higher areas of salt marsh preferred by both cattle and nesting redshanks during the sensitive times.

There are many other flocking birds. In the United States, as October fades into November, we have flocks of Red-winged Blackbirds and Common Grackles foraging on leftover grain. European Starlings flock by the thousands and form mesmerizing murmurations as they descend to roost. Nearly all species of shorebirds form foraging flocks.

House Finches flock in winter, arriving at my feeders in groups of twenty or so. Across the globe in the Great Rift Valley lakes of Kenya and Tanzania, Lesser Flamingos forage in flocks of hundreds of thousands.[12] The groups are so common that we have developed names for them. You might have heard of a murder of crows and a gaggle of geese, but how about a parliament of eagles, a bouquet of avocets, a scold of jays, a quiver of hummingbirds, a covey of quails, or a round of robins?

Scientists spend their days studying individual species at promising locations. When there are enough such studies, others amass them to look for general patterns across all the species, an approach called a meta-analysis. A master of this method is a scientist who lives near Montreal, Canada, Guy Beauchamp. He received his doctoral degree at the University of Cambridge in the UK studying under Alex Kacelnik, who is famous for work on avian cognition, particularly in New Caledonian Crows. Beauchamp settled into a job in his native Montreal as the statistician for a veterinary college, and his research on birds became his hobby—and a productive one at that. He wrote three books. This was possible because he took a different perspective. He did not do fieldwork himself but instead gathered the information collected by others so he could discover general patterns.

Beauchamp conducted a meta-analysis on flocking. In already published works by others, he found data for 1,231 species in 153 different families of birds.[13] Of these, 84 percent of the families had at least one species that flocked. Beauchamp tried to figure out from these data what characteristics of the birds made them more likely to forage in flocks. First, he was able to rule out nocturnal species; they tended to forage independently. Next, he considered diet and found that flocking was more common in birds with plant rather than animal diets. Birds like Red-winged Blackbirds flock extensively in the nonbreeding season when their diet is plant-based. Beauchamp found that species that foraged in the air were more likely to flock than terrestrial or aquatic species, probably because air predators go after aggregated prey like insect swarms and bird flocks. Those grouping species that did feed aquatically were more likely to be at sea or on the shoreline, like the redshanks.

Another variable Beauchamp studied was the environmental features that might have an effect on flocking. For example, one could compare islands with the mainland. On islands that are distant from the mainland, like the Galápagos off the coast of Ecuador, there are so few predators that some birds have even lost the ability to fly. Indeed, since there are generally fewer predators on islands, and since we know that flocking is mostly about avoiding predators, a researcher could predict that all else being equal, there would be fewer foraging flocks on islands. Beauchamp wondered this very thing, knowing that this sensible prediction might not hold up if the tendency to flock were fixed in the birds' nature (they wouldn't lose the ability to flock all at once, just as we wouldn't suddenly grow a sixth finger even if it were advantageous).[14]

Beauchamp identified twenty-two islands that lacked predators of birds during the nonbreeding season, when they are likely to form flocks. Generally, these were tropical islands far from any mainland, making it hard for predators to cross over. Then he looked for pairs of bird species that had a member on the mainland and a closely related species on the island. He predicted that the island species would have smaller flocks, or a lesser tendency to flock, than those on the mainland. If this hypothesis held up, it would be a natural test of the idea that flocks form to avoid predators.

As predicted, Beauchamp found that flock sizes on the islands were smaller than flock sizes on the mainland. He was able to rule out the notion that less food or fewer birds caused this pattern, making it likely that the lack of predators led to smaller flock sizes on islands.

Of course, as we've already established, birds flock not only when they are foraging. They also flock as they fly, both locally and in long migrations. A clear example of flocking during flights close to home is homing pigeons. They are a domesticated subspecies of Rock Pigeons bred and named for their famous ability to navigate back to their home roost no matter where they are released. That is, they come home if they are not seduced by a Horseman Thief Pouter, a pigeon trained to lure pigeons of the opposite sex from other roosts back to his own. Homing pigeons' uncanny navigational talents once made them crucial as messengers in times of war, but they also come in handy for researchers eager to study their ability to avoid predators when in a flock.[15]

Daniel Sankey and his group took advantage of homing pigeons' return flights by fitting them with GPS tags and watching differently sized flocks respond to a predator carrying a tag of its own. Through GPS, researchers could track exactly where all the birds were at any given time.

The pigeons were caged and fed at the Royal Holloway, University of London, about 23 miles (37 kilometers) west of London. For the experiments, the birds were taken to a nearby natural area, Chobham Common, and released. Through GPS, the researchers tracked their 3-mile (4.8-kilometer) journey home. Additionally, they measured changes in the birds' weights to analyze how much stress they were under (weight has been shown to be correlated with the stress hormone corticosterone, and measuring weight is easy on the bird and for the researcher). The point of the experiment was to see if the birds behaved differently when the researchers exposed them to a predator.

The predator in question was a robotic falcon, aptly named RobotFalcon, built by Robert Musters to mimic a male Peregrine Falcon in both appearance and behavior. RobotFalcon was piloted using a live camera feed connected to video goggles, giving Musters a bird's-eye view of the pigeon flock as it returned to its roost. Musters imitated falcon flight, sending RobotFalcon soaring up before diving nearly straight down on his targets. RobotFalcon looked like a predator, acted like a predator, and could approach when and where the researchers wished, with the added advantage of running on batteries instead of on pigeon flesh.

For each trial, the researchers gathered up the pigeons in a covered wicker basket, took them to Chobham Common, and released them. Sometimes they let the birds fly home unaccosted by RobotFalcon, but sometimes it was lying in wait. Crucial to the whole study were the GPS loggers, far fancier than the GPS on your cell phone. They tell the position, trajectory, and speed of each bird and record this information five times a second.

So how do pigeon flocks react to predators? If, like the Common Redshanks, the pigeon flocks were primarily organized according to each bird's desire to hide behind the others, the researchers expected the flocks should cluster more tightly when RobotFalcon appeared.

The GPS data told a different story. The researchers found that when the pigeons noticed RobotFalcon, they turned sharply away from it rather than toward the center of the flying group, as might be predicted by Hamilton's Geometry for the Selfish Herd. Furthermore, it happened that the pigeons paid attention to movements of their neighbors as well as the predator. If their neighbor swerved, the pigeons swerved, too. They didn't wait to see *why* their neighbor swerved, which made the researchers conclude that the birds were responding to the birds closest to them, who might have seen the predator already.

These data suggest that in highly aligned flocks, moving toward the center is not as advantageous as moving in the same direction as your neighbor. This fits with the sort of movement that fish shoals make, according to James Herbert-Read.[16] The result is a kind of wave as individuals responding to the movements of their neighbors turn away from the threat.

Other researchers found that pigeons actually pay a cost for flying too close together. James Usherwood and his team found that pigeons in a flock took more, shallower wingbeats, requiring more energy than when they flew alone.[17] This kind of flight allowed them more control, so they would not inadvertently bump into another bird. The benefits of avoiding predators must be high enough to make it worth flying this way.

But birds that fly together do not always spend more energy the way the pigeons do. After all, pigeons are not flying far. Other birds can fly thousands of miles on their migratory flights. They need to conserve their energy. One of the most spectacular forms of flock movement is the famous V formation, wherein each bird follows the one in front, just slightly offset to catch the updraft, similar to cyclists drafting in a peloton.

Henri Weimerskirch compared Great White Pelicans flying in groups in V formations with birds flying alone to see whether they saved energy.[18] He and his colleagues put electronic heart-rate monitors on eight Great White Pelicans trained to fly after a motorboat and an ultralight airplane in Djoudj National Bird Sancutary, Senegal. The heart rate of the birds in the V formation was about 13 percent lower than that of the birds flying alone. By positioning themselves behind another bird's wing tips, the pelicans could take advantage of an upwash of high-pressure air left in its wake, letting them flap less and save energy. This is a substantial advantage during migration. It's a reason to fly together.

What about the bird at the tip of the V? It is time to turn to another large migratory species, the highly endangered Northern Bald Ibis. If ever there were a bird flying out of Mordor, the darkest place in *Lord of the Rings*, it might be this one, its black wings beating slowly behind a bald head and long pink bill curved dramatically downward. Find it on the ground in breeding season, and you will see its feathers sticking straight up, the ultimate punk hairdo. Ancient Egyptians revered this bird and considered it to be divine, which is why it appears in hieroglyphs.[19]

Bernhard Voelkl and his team looked at how Northern Bald Ibises shared the front point position—the spot at the tip of the V.[20] They put data loggers on fourteen juvenile Northern Bald Ibises that had been raised at the Salzburg Zoo as part of a conservation effort to repatriate the birds to the wild. Since the young birds needed to learn the migration route, and since the investigators wanted to entice them onto a specific arc from Salzburg, Austria, to Orbetello, Italy, Voelkl's group trained them to follow an ultralight paraplane.

So, what did the researchers learn about who follows whom? I would have guessed that the ibises would mostly follow a particular bird in front of them. But the pattern was a lot more dynamic. Individuals were behind another bird in that favorable updraft position about a third of the time, but they changed positions frequently, following one bird for mere seconds before moving to follow another. This led the researchers to ask how often a leading bird dropped back to follow the bird that had just been behind it. It turned out, it happened more often than randomness would suggest—a *lot* more often. Voelkl turned to the theory of reciprocal altruism to explain this behavior. Essentially, there is an advantage to following and a cost to leading. If certain birds only followed, then the leaders would be taken advantage of. Northern Bald Ibises apparently avoid this problem by switching roles constantly.

Of course, Northern Bald Ibises are only one species. The frequent switching Voelkl observed is not what I see along the Texas coast, where Brown Pelicans cruise along, keeping flapping to a minimum as they glide low above the water—their leader keeps its position for minutes. And it is not what I see in Snow Geese as they fly from the Mississippi riverlands north to their breeding grounds—the Snow Geese typically have one bird in the front position for as long as I can see their V. Presumably there are positional trade-offs for these birds, too. These observations make me wonder whether Voelkl's young ibises might have formed larger strings of followers had they learned how to migrate from adult ibises, rather than having to figure it out from following an ultralight paraplane, or whether their patterns might change as they age. Either way, it is gratifying to see captive-reared birds succeed in learning to migrate through Voelkl's effort and return to some of their former habitat.

Birds fly efficiently because they must. They have long distances to travel, winging their tiny bodies north to vacant nesting locations in spring and south to lush warmth in winter. Migrations are something to witness. We can see the heart-stopping great wildebeest migrations in East Africa because they are on land, but birds turn the whole world into a great migratory palette, painting the globe with their colors as they move with the seasons.

There are advantages to flock migration besides energetic efficiency. A group might better figure out the precise direction to fly, for instance; this is particularly advantageous for young birds on their first flight south. They might also be better at connecting with the thermal air currents that lift them high and carry them without effort. When they stop during migration to feed and rest, a group also has a ready-made flock for protection and food discovery.

Hawks, though they typically migrate in flocks, do not form V's. This is because they depend on the movement of the sun-heated air to travel. They ride thermals and other air currents up and then glide to the next thermal. They funnel into tighter groups as they approach geographic features, like hills and ridges, that favor this. Emily Shepard explains that birds like the migrating Broad-winged Hawks gain altitude in thermal updrafts, in which they circle to stay in the most strongly lifting airs.[21] Then, as they leave the updrafts, they glide—as far as 16 feet (5 meters) for every foot that they climbed. The hawks, Shepard writes, are also experts in knowing where the air rises. Since warm air rises and cold air sinks, when the air hits a hill, these forces create an updraft that the birds can use to rise high. The wind currents are powered by the warming sun, so they are daytime features largely unavailable to nocturnal migrants like songbirds.

The hawks that I have visited most have been at Hawk Watch sites, where they are counted by researchers and volunteers. Smith Point, Texas, and Whitefish Point, Michigan, are places where land features reaching into water funnel the hawks. Hazel Bazemore County Park near Corpus Christi, Texas, is another famous Hawk Watch site. Paul Kerlinger and Sidney Gauthreaux Jr. quantified the migratory flights of Broad-winged Hawks during spring migration, focusing on an area a little farther south of Hazel Bazemore, the Santa Ana National Wildlife Refuge, right on the border of Mexico.[22] I visited Santa Ana with my husband in April 2023, nearly at the end of the hawk migration north, though I could easily see South Texas specialties like Great Kiskadees and Green Jays. We stood on the bank of the Rio Grande and looked across its green waters into a tangled thicket—Mexico. On my prior visit when I stood on this bank, looking into this same thicket, a Green Kingfisher flew out, clattered its alarm, and then flew into the United States, perching on a dead limb over the water watching for fish.

Kerlinger and Gauthreaux had what they called the Avian Migration Mobile Research Laboratory, a 23-foot (7-meter) motor home with radar and other equipment (including their trusty notebooks and binoculars) to monitor hawk migration. They coordinated the information on flocks coming from the radar with their actual observations of flock sizes and behaviors. Hawks migrate during the day, when they can ride thermals, and descend at the end of the day to roost in the park. Winds were generally to the south, and when they were not, the birds did not fly. In Kerlinger and Gauthreaux's twenty days of observation, nearly eighty-five thousand Broad-winged Hawks crossed the Santa Ana National Wildlife Refuge.

The team noted that flight patterns were regular and involved the hawks soaring high with the hot air and then gliding with little flapping. By 11 a.m., the birds gathered into distinct flocks of more than one hundred birds. Some flocks of more than one thousand birds spread long, rather than wide, as they glided from one thermal to another. The birds could stretch half a mile apart, increasing the chance that some would encounter another thermal and then soar, the others following them upward.

Wake up to flocks and you will start to see them everywhere, from the ducks in a city pond to the chickadees or tits hunting for insects deep in winter. From now on, I hope you'll look at them just a little bit longer and with a little bit more wonder.

CHAPTER 2

Communal Roosts

Why Sleep Together?

My son told me we needed to visit Cahokia Mounds in Illinois if we wanted to see the blackbirds. We visited this pre-Columbian city (which peaked in population around 1100 CE) at dusk, when the huge earthen pyramids loomed against the darkening sky. We crossed the highway from the highest pyramid and looked at the trees edging the eastern meadow around the visitors' center. As promised, they teemed with Red-winged Blackbirds.

It was far more than hundreds, even more than thousands. I could not imagine how to count the birds or to tell how many Common Grackles, Rusty Blackbirds, and European Starlings were also part of the roost. The birds flew in and out, each vocalizing. I tried to look at one branch and watch as the birds on it jostled for position. But other branches distracted me. Sometimes all the birds lifted up, as if an electric shock had passed through the naked trees. Then they quickly settled. More birds flew toward the sunset and away from the sunset. When new arrivals came, the birds lifted and adjusted. By the time they settled, it was too hard to see in the darkening night.

If you spy birds sitting together in a tree in the evening, you have probably come across a communal roost. For a place to be a communal roost, the birds must sleep there. Birds sleep in ways similar to mammals like us, with rapid eye movement (REM) and slow-wave sleep (SWS) periods, but only half of their brains sleep at a time.[1] This half-brained sleep helps protect them from predators and allows them to sleep and fly at the same time. Importantly, roosts are not nesting areas (birds that nest in a group are called a colony, which I'll discuss in chapter 4), so they are most common outside of the breeding season. Roosts can be as small as a couple of Bushtits on a bare branch, or as large as millions of quelea carpeting many adjoining trees.

Communal roosting is a common behavior that should be easy to understand: It's one of the few times when birds hold still. This means they can be counted, watched, and studied. However, we know there are costs to communal roosting. In a communal roost, birds can transmit and catch infectious diseases or ectoparasites (external parasites like ticks) more easily. Birds in communal roosts will also be subject to other ills of group living, like squabbles, fecal rain falling on lower birds in a roost, and competition for a good spot. So why do they do it?

One of the first comprehensive studies on communal roosting was done by Arthur Allen, who started Cornell University's famous Lab of Ornithology.[2] This is the group that now brings us ways to follow birds using eBird and Merlin, free apps for your smart phone. Allen studied Red-winged Blackbirds at a place known as Renwick Marsh at the head of Cayuga Lake near Ithaca, New York. He said:

> *The first flocks to arrive in the evening . . . are seen flying lower and lower . . . and are intent upon but one thing—the finding of a place to roost. The place selected is a spot where the flags [cattails] are not quite so completely burned and a little more shelter is afforded. Toward this spot, as if with some previous knowledge of its location, all of these later flocks direct their flight, and disappear into the cat-tails. . . . Every available perch, not so high as to be conspicuous, is filled with birds down to the water's surface, but were it not for the unspeakable din that arises from the hundreds of throats, one*

> *would scarcely be aware of their presence, so inconspicuous are they against the dark water.*

Allen goes on to describe how the birds wake and leave in the morning as dawn lights the marsh.

Though Renwick Marsh has since been destroyed, converted into a dry sandflat due to dredging of the sandbar that protected it, Red-winged Blackbirds are still going strong. They are the most common of all the North American communal roosting birds and among the most common birds in America.[3] They can be found breeding anywhere there is a scrap of wetland, a little bit of marsh, or even a drainage ditch. They are despised by farmers because they eat grain, especially as they forage from their fall and winter roosts, where they congregate in millions, covering the bare limbs of every tree.

Patrick Weatherhead and Drew Hoysak looked into roost structuring in a small roost of about two hundred male Red-winged Blackbirds in Mooney's Bay Park in Ottawa, Ontario, and found that older males got better positions on the most desirable cattails because they were dominant over younger males.[4] The best positions tended to be higher ones, where the birds would not be targets of the excrement of others, but the birds would quickly move lower—displacing the subordinate birds—if a predator was sighted.

Estimating the size of enormous groups is not easy. Brooke Meanley described roosts in the southeastern United States from 1950 to 1964, writing that they ranged from "small," composed of a paltry twenty thousand birds, up to massive winter roosts of up to ten million.[5] Different species roost together: Red-winged Blackbirds frequently roost with Common Grackles, Brown-headed Cowbirds, Rusty Blackbirds, and European Starlings. In a densely wooded Arkansas roost, Meanley found that European Starlings rested highest in the trees, followed by Common Grackles and male Red-winged Blackbirds, then Brown-headed Cowbirds and female Red-winged Blackbirds, with the lowest birds being more female Redwings and Rusty Blackbirds. He observed other roosts with various kinds of stratification by species and sex.

Meanley further observed that the birds would roost together even if it meant flying as many as 35 miles (56 kilometers) away from forage sites. Imagine if your own bed were 35 miles from your breakfast table! Roosting together must provide strong advantages if birds are willing to go so far out of their way to do it.

There are three main hypotheses for communal roosting. The first is that birds may stay warmer in cold weather when they are snuggled together. Birds are warm-blooded and need to maintain their body temperature, no matter how frigid the night. Huddling with others can reduce individual energy costs, and so may be especially important for smaller birds living in colder climates.

Take, for example, Bushtits, which are so tiny that an adult weighs only 5 grams, about as much as five shelled almonds. They are found from southern British Columbia in Canada through the mountains of Mexico, and through much of the western United States. I last saw them while visiting my daughter in Grand Junction, Colorado, in chilly December. The nearby McInnis Canyon National Conservation Area is beautiful, composed of reddish sandstone weathered into fabulous shapes, pillars, and outcrops. I watched as the Bushtits flitted into juniper bushes bordering a high trail. They were constantly on the move and incessantly vocalizing to keep in touch.

Susan Chaplin investigated the energetic lives of Bushtits while she was at Occidental College in Los Angeles.[6] She temporarily brought six birds into the laboratory to study how their metabolism changed at different temperatures and in different group sizes. Chaplin found that Bushtits ate 80 percent of their weight in mealworms every single day. If forced to sleep alone, a single bird lost 10 percent of its body weight every night, just to maintain its body temperature. The Bushtits she allowed to roost in groups, however, used significantly less energy overnight at all temperatures. Even just a single roostmate was enough to lower a bird's weight loss to 8 percent of its mass, a 20 percent lower nighttime cost.

Long-tailed Tits are close relatives of Bushtits. I often saw Long-tailed Tits on the tree-lined streets of Grunewald in Berlin, Germany, near the Wissenschaftskolleg, or Institute for Advanced Study, where I spent a year just before the pandemic. I began each day by watching the Long-tailed Tits and returned to my desk enriched. One winter day, I watched a group of about six birds as they foraged in the trees, chattering constantly and moving quickly just as the Bushtits did in Colorado.

Ben Hatchwell and his team looked at how Long-tailed Tits survived cold nights in Melton Wood, near his university in Sheffield, UK.[7] Like Chaplin, Hatchwell collected flocks into temporary captivity in large outdoor aviaries, where they were fed wax moths, mealworms, and crickets. Hatchwell and his colleagues videotaped the individually tagged birds, weighing them before nighttime roosting and at dawn the next day, after a night of communal roosting.

Like the Bushtits, the Long-tailed Tits lost about 9 percent of their total body mass overnight. Birds on the ends of the roosting line lost more than those insulated by a neighbor on each side.

Andrew McGowan led another study on the same captive flocks to figure out how they sorted their positions in a roost.[8] After the flocks got used to the outdoor aviary, McGowan began videotaping them with infrared night vision. Birds jostled for the best positions as the night began but then generally did not change position until dawn. They may not move much once they have settled because such actions would make everyone colder. Movement could also draw the attention of night predators to the group.

McGowan and the team also watched and videotaped the birds during the day to determine whether there were dominance hierarchies that might explain which birds got the best positions. They looked for tussles over food items, displacement, and conflicts over roost positions, usually taking the form of pecks by dominant birds on the napes of their victims. Four of the flocks did not show dominance behavior and were quite peaceable. The remaining flocks pecked, tussled, and displaced each other enough for the researchers to put them in a hierarchy from the most aggressive bird to the most subordinate. The researchers were interested in whether these behavioral actions would predict where the birds ended up in the nighttime roost. In four of the seven flocks, the same two individuals ended up on the edges every night. The birds in the inner positions seemed more randomly distributed. Most of the time, the most subordinate individuals occupied the chilly end positions.

Long-tailed Tits are not the only birds that use dominance to decide preferred roost positions. Bronze Manakins,

European Starlings, and Rooks do, too. Other species do the opposite, using dominant individuals as bookends to a communal roost. One example of the latter is the Varied Sittella, a tiny, common Australian bird that lives in dry eucalyptus, acacia, or casuarina forests.[9] As they forage for insects, they chatter constantly, just like Bushtits. A breeding pair usually has helpers—young from previous broods or an unrelated male that joins the group hoping for future reproductive opportunities. At night, the social groups roost, keeping warm by sitting in close contact, all facing the same way, as is generally true for birds that roost for warmth. The mother of the young is in the middle of the group, and the father and another male are at the ends.[10] The juveniles tuck in somewhere in the middle and are the first to close their eyes, just as we wish our own children would do. The male at the position toward the outside of the tree, with the best view, stays awake the longest. Richard Noske suggests that the bird is alert to danger. Why should the dominant male have the worst position? Unlike Bushtits, Varied Sittellas roost with their family members, and the male takes the worst spot as a way of protecting his young.

The second reason birds might roost together is for the same reason they flock together—predators. Sleeping together at night, just as foraging together during the day, can reduce predation on any one individual.

A vivid example of roosting under predator threat comes from the work of Rob Bijlsma and Bennie van den Brink on Barn Swallow roosts in central Africa.[11] Like those of the Red-winged Blackbirds, Barn Swallow roosts are enormous, and they are always watched by predators. Bijlsma and Van den Brink worked at a site called Boje-Ebok in the foothills of Afi Mountain in Nigeria, an area with primary rainforest on the slopes and agriculture in the valleys. The Barn Swallows roosted in fields of elephant grass (so named because it can grow to 15 feet [4.6 meters] tall or more). While the investigators were watching the roost, so were the predators—Bijlsma and Van den Brink counted seven African Hobbies, three Yellow-billed Kites, a male Eurasian Marsh Harrier, a pair of African Harrier-hawks, two Red-breasted Sparrowhawks, a Shikra, a Lanner Falcon, a Barn Owl, two Eurasian Scops-owls, a pair of African Wood Owls, and a Senegal Coucal all keeping their eyes on the study site.

Bijlsma and Van den Brink observed the swallows in the early mornings as they flew away from the roost and at twilight when they returned—the most dangerous moments of the day for many roosting birds, because the concentrations of birds draw the attention of predators. Sometimes the birds came in what the researchers called "an endless stream," making counting impossible. They describe the morning departure thus:

> *Swallows started bill-snapping and twittering. In the following minutes the tension built up and singing intensified in strength, an enchanting moment in a rainforest setting. Many Swallows started preening, wing-stretching and fluttering. The first Swallows normally departed from the roost between 0 and 9 minutes before sunrise. . . . In general, Swallows exhibited two ways of departing, i.e. (a) "rolling downhill" in fast flight, hugging the elephant grass or using openings in the grass to fly even lower, and (b) flying straight up, as fast and as steep as possible, sometimes using tight spiraling to gain height.*

Nearly all the birds flew straight up; the rolling-downhill behavior occurred only after falcons appeared.

The evening return flight was similar—it happened fast and all at once, with nearly every bird completing the landing within the first nine minutes after sunset. The language of the researchers captures what it was like to be there: "Arriving birds flew very high, almost invisible against the darkening sky. The descent was spectacular. Dense throngs of Swallows 'rained' downwards, diving into the roost at great speed. Within a couple of minutes, hundreds of thousands of Swallows had entered the roost."

African Hobbies clearly anticipated the departure and arrival of the swallows, because they came to the roost before the swallows left in the morning or arrived in the evening. Hobbies hunted alone or in pairs, flying low and chasing birds or diving on them over and over again. In the evening they attacked from high up, causing the swallows to circle even higher until they all plummeted down into the grass "like an inverted tornado." Thirty-eight percent of all African Hobby attacks were successful during the time that Bijlsma and Van den Brink watched, an unusually high success rate that they attributed partly to the site, a large clearing in a forest. Attack success rates of African Hobbies in the Ivory Coast and Ghana were 25 percent of all attacks.

What was the result of all these attacks? The researchers estimated that African Hobbies killed about fourteen Barn Swallows a day, totaling about twenty-five hundred over the six months the swallows were at the roost. Other predatory birds claimed about double that number, though even all combined, the casualties were few compared with the huge number of Barn Swallows at the roost.

Though roosts attract predators, the roost is nonetheless

likely the safest place for any given Barn Swallow to be. Bijlsma and Van den Brink found that African Hobbies were most successful at predation when they could target small flocks of fewer than fifty Barn Swallows, probably because it was easier to follow a single bird to a successful capture. Being part of a roost of millions, all arriving and leaving at various times, helps the Barn Swallows confuse the predators and lowers the risk to any one bird.

Roosts—like the Barn Swallow roost at Boje-Ebok, which has been there for decades—tend to be used over and over again. This makes them a predictable resource for predators of all kinds, including humans. Locals have, for instance, identified the Boje-Ebok Barn Swallows as a source of food and sales, and they harvest an estimated 105,000 to 462,000 birds per year. People ambush the birds with a unique tool: a yam on a stick covered with twigs and coated in an adhesive palm resin. Though Barn Swallows are not rare and occur through most of the world, the European population of the birds has declined. We do not know how much the Barn Swallows' treatment in their wintering grounds in Africa has contributed to that decline, compared to changes in Europe, nor can I find information on how the Boje-Ebok Barn Swallows are faring now. But I do know that wherever humans suffer from hunger, wildlife suffers, too. These issues need to be addressed together.

The third proposed reason for communal roosting has to do with finding food. There are two basic ways in which communal roosting could facilitate foraging. First, birds can watch other birds and follow those that are finding food. This is called the "local enhancement hypothesis." I don't love the name, because "local enhancement" sounds vague, but apparently it comes from bird behavior inadvertently enhancing the environment for others by revealing food sources. Under the local enhancement hypothesis, when birds leave a roost in the morning, they spread out over the landscape looking for food. Birds that find food behave in ways that can be seen by others, who will then join them. For example, Cedar Waxwings fly up from their nighttime roosts and look for trees where the fruit is ripe enough to eat. Those that find the right trees settle down to eat. Other Cedar Waxwings join them. Luckily, there is plenty of fruit for all.

The second reason that communal roosting might facilitate food finding is called the "information center hypothesis." This hypothesis postulates that there is active communication at the roost, in which birds that found food the day before tell others where to find it the next day.[12] It is tough to imagine the conditions under which this would work. Why, for example, would a bird that knows where the food is even go to a roost and tell others unless they are relatives?

Ever since Peter Ward and Amotz Zahavi espoused the information center hypothesis, others have made the effort to dissect their central argument to see what kinds of avian capabilities it would require. A comprehensive report came in 1987, a full fifteen years after Ward and Zahavi published their controversial paper.[13] Doug Mock, Tim Lamey, and Desmond Thompson laid out seven expectations that would need to be met for the hypothesis to make sense. First, birds should return to a good feeding site. If they don't return but instead hunt for a new feeding patch the next day, then there is no point in following yesterday's great foragers—they would not have any special knowledge. Second, there must be variation in the foraging success of roost members. No point in following someone who's no better than oneself at finding food. Third, the unsuccessful birds must be able to figure out who has been successful at foraging the previous day. There is no particular reason why successful birds, sharing their roost with competitors, should advertise their successes the way honeybees do with their dance language among family. Fourth, successful *and* unsuccessful birds should leave the colony at the same time, so that the latter can follow the former. It is often true that birds leave roosts in a big push right after sunrise. Fifth, some birds must actually follow others as they leave the colony. This may be difficult to assess if the birds generally fly in the same direction on leaving a colony. Sixth, the formerly unsuccessful birds, on reaching the better food patch, must be allowed to actually feed and not be chased away by those they followed. Seventh, following must actually benefit the birds that follow.

Mock and his team did not believe these seven conditions were very likely, but they nevertheless explored their likelihood in Great Blue Herons and Great Egrets. Their research failed to provide evidence for information centers. Birds that were poor at locating fish, their primary food, did not follow successful birds. Low-success birds did not stay at the colony longer.

Mock had presented the necessary conditions for an information center in a way that could be used to predict when they would work. However, what if there were ephemeral, large sources of food that were hard to discover but, once found, provided food for all for a few days? This would describe the practices of a common carrion feeder, the Black Vulture. The Black Vulture often follows another species of vulture with a superior nose for carrion, the Turkey Vulture, in order to find food.[14]

Back in Texas when I was teaching bird field biology at Rice University, we came upon a roost of about seventy Black Vultures. It was not entirely clear if it was a true roost or a feeding ground, since the birds were perched in an area where the remains of catfish were piled high. I'm inclined to think it was a roost since the birds were not feeding; instead, they were resting on rusted and abandoned tractors and other equipment and in the trees nearby.

Black Vultures are not the most attractive birds. Their heads are naked, so they can reach deep into a carcass without despoiling their feathers. They cool themselves by excreting on their bare legs. This means that researchers cannot band them around the leg the way they do with other species. Instead, they carefully put cattle ear tags on their wings.[15]

Patricia (Patty) Parker, recently retired from the University of Missouri—St. Louis, where she was the E. Desmond Lee Endowed Professor in Zoological Studies, is the Black Vulture roosting expert.[16] From her work, I could see that vulture roosts might serve as information centers in a way quite different from that envisioned by Ward and Zahavi. But to understand this, it is worth knowing a little more about how Black Vultures live and why they need roosts.

Black Vultures used to nest in large hollow logs and caves. As humans changed their landscape, the birds adapted to what they could find. Often, that was abandoned buildings. Each pair found their own place to nest, and the female laid her eggs on the floor. These pairs were entirely true to each other, with no indication that they ever mated with anyone other than their partner.[17] Parker found these nests at her study site in Chatham County, North Carolina, not far from where she was a grad student at the University of North Carolina at Chapel Hill.

The rural landscape where Parker worked was worlds apart from the ivy-covered buildings of the university. When Parker did her research in the early 1980s, she was surrounded by independent chicken and pig farms run by families who had farmed the land for generations. Since Parker had spent her childhood from the age of ten in North Carolina, she sounded and acted in ways that put these farmers at ease. They trusted her with their vultures—and often with their secrets. She was even allowed to rent an old hunting cabin so that she could be close to her vultures. The cabin had minimal electricity and no heat or plumbing, so Parker's stepfather helped dig an outhouse. She added a woodburning stove for warmth. When it wasn't warm enough to bathe in the pond out back, she heated bathing water on the woodstove in an old, blue-flecked enamel basin. My own grad student housing in Austin, Texas, was basic, but not *that* basic!

A particular feature of this North Carolina landscape

made abandoned homes particularly common. When paved roads were put in during the middle of the last century, they did not follow the original dirt roads; instead, they ran parallel, a few hundred yards away. The next generation built more modern homes along these new roads, leaving their parents' and grandparents' homes to become covered in kudzu and other vegetation. In those ancestral homes, the Black Vultures found places to nest.

Even though abandoned homes and sheds were comparatively common, they were not necessarily close to food supplies for Black Vultures. These birds eat dead animals, but not *long*-dead animals. By four days a carcass is past the stage a vulture will eat it. It is an ephemeral resource that, when found, is usually abundant enough to feed all the vultures. Might a communal roost be just what Black Vultures need to find the next dead animal?

Another feature of Black Vulture life makes a food-finding hypothesis for Black Vultures even more likely: Parents do not trade off nest and nestling care very often, only about once a day, in the late morning. So the parent relieved of nest duties emerges into a world without any idea of where to find carrion. Instead of flying and looking around on their own, they first fly to the roost. In the area where Parker worked, there were seven Black Vulture roosts that, according to local farmers, had been used for decades.

To see exactly what the connection between nests, roosts, and prey was, Parker had to individually identify the birds first. She needed to be able to recognize which birds came from which nest and where the birds went. To do this she put cattle tags on their wings with colors and numbers that could be read from a distance. In addition to tagging adults, Parker tagged their young while they were still at their nests so she could identify families. She was able to do this time-consuming task on 344 vultures out of a population of about 1,200.[18]

Black Vultures not tending young sleep at the roosts. There is no indication that roosts serve either for warmth or for protection from predation for these large birds. The other advantage to roosting is that it facilitates foraging. Do Black Vultures follow others as they leave the roost in ways posited by information center theory? Or do they look for where other vultures are, according to local enhancement theory?

Parker watched thirteen carcasses she had hidden to see if more vultures showed up the second day than the first. If this was the case, she argued, the extra birds would have followed those that had found the carcass on the first day from the roost. She found that there were more second-day arrivals, supporting the information center hypothesis.

In the second experiment, Parker captured thirty-three Black Vultures and released them at a roost after they had been away from food sources for more than two and a half days, longer than a good food source lasts.[19] These birds would not know about good food sources. As the birds left the roost, twenty-nine of thirty-three were in the rear halves of their groups, as expected if they were following others to carcasses, again supporting the information center hypothesis.

Clearly, Black Vultures get a lot out of roosting together. They would not know where carrion had been found if they just came off nest duty without a roost to visit. The birds that were most likely to know where food was were the older birds, who often have young or other relatives at the nest. Parker discovered that families stuck together at the roosts.[20] Mates were together at the roost when they were not actively sharing nest-parenting duties. Their fledged young were also with them. They even socialized more with neighbors in the study area who might have been more distant relatives.

Another bird whose roosting tendencies are hard to explain is the bird that's on the seal of the United States of America: the Bald Eagle. When Bald Eagles are not breeding, they typically spend the night in large roosts that can cover acres. The eagles prefer locations with very tall trees emerging above the canopy, ideally away from humans and near food sources.[21] They are large birds with no natural predators, and it's unlikely they need to huddle for warmth (indeed, they do not roost close enough together for warmth). Their loose roosts tend to be along rivers or lakes, which provide the fish Bald Eagles love. During the day, they spread out along the water body to fish alone. So, what might the roosts be for? Perhaps roosting has yet another purpose.

In 1983, when Bald Eagles were still listed as endangered in forty-three states and threatened in another six, George Keister Jr. and Robert Anthony did a survey of their communal roosts in the Klamath Basin of Oregon and California.[22] As many as five hundred eagles wintered there, feeding on fish in lakes and wetlands. Near the feeding areas, the investigators identified five different communal roosts, each covering dozens of acres. The roosts were called Bear Valley, Caldwell, Cougar, Three Sisters, and Mount Dome. The roost trees tended to be older Douglas firs and ponderosa pines that were taller than surrounding trees. The Bald Eagles flew as far as 5 miles (8 kilometers) to be in these preferred large trees. Maybe a scarcity of large trees is another reason for communal roosts.

Another possibility is that the eagles gain social information or even find mates at roosts. Amy Yackel Adams and her team studied interactions among eagles in a roost on the north fork of the Nooksack River in Washington State.[23] They were interested in interactions between the eagles. The most straightforward interaction was when one eagle flew so close to another that the first eagle moved away. Eagles changed positions in the roost more in the evening than in the morning, they found; no surprise since in the morning the birds were preparing to leave. Also, most of the birds that relocated in the evening were immature eagles generally moving closer to other immature birds rather than toward older birds, so maybe they were looking for mates.

Another pattern the researchers observed was that the eagles were more likely to move from edge trees to central trees in the roost, particularly when the temperature was below freezing. They may have been trying to get to a warmer microclimate more sheltered from wind. However, they did not move close enough together to share warmth, the way the tiny Bushtits do.

Taken together, the most likely reasons for the roosts of Bald Eagles are the paucity of tall trees and the possibility that roosts facilitate social interactions among immature birds. But whatever the reasons, the good news is that Bald Eagles have survived a near mortal blow from human-induced extinction with DDT pesticides preceded by excessive hunting. They are now abundant enough that they *can* form roosts. According to *Birds of the World*, we now have about 316,700 Bald Eagles in the contiguous United States and another 58,000 from British Columbia to Alaska.[24] Since 2007, they are no longer considered endangered. Let's keep it that way.

Instead of looking at individual species, Guy Beauchamp took a phylogenetic approach to the puzzle of communal roosts.[25] He looked at the avian tree of life, which shows the evolutionary relationships among bird species. It is like a family tree. More closely related species are connected with shorter branches. Ultimately, all the birds connect 150 to 165 million years ago, when birds first arose from a theropod dinosaur.[26] We do not have ancestral birds around anymore, but we can still trace ancestral traits using their modern descendants. Take a species of bird that roosts communally, along with all its closest roosting relatives, and follow that lineage back to where it connects to non-roosting birds on the tree of life. We can then see what distinguished roosting and non-roosting bird species.

Beauchamp looked for cases where communal roosting behavior arose from an ancestor that did not roost communally. He also looked for cases where the ancestor had communal roosting and the descendant did not. He considered these latter cases to be losses of communal roosting. Out of 201 branches on the avian tree of life, he found communal roosting arose in 21. The biggest predictor of the origin of communal roosting was foraging in flocks.

Loss of communal roosting happened in 32 branches on the tree. The ones that lost communal roosting were also more likely to have lost flocking behavior. Thus, it seems clear that birds that feed together during the day are more likely to roost together at night. If only we could understand bird chatter, we might better understand communal roosts.

CHAPTER 3

Mixed-Species Flocks

Follow the Alarm Caller

Step into the nearest forest, anywhere on the globe. Listen. You might have to wait or walk a few steps. What do you hear? On my late-summer evening walk in northern Michigan, it was Black-capped Chickadees and Tufted Titmice twittering along with the occasional White-breasted Nuthatch or Downy Woodpecker. If I had been in the forests along the Tambopata River in Peru, I might have heard Cinereous Antshrikes, Bluish-slate Antshrikes, and a Gray Antwren. If I were in the Fushan Experimental Forest in Taiwan, I might have heard Grey-cheeked Fulvettas and Yellow-throated Minivets. Mixed-species flocks are everywhere.

It was six in the morning when I first became aware of the magic of mixed-species flocks. My family and Claudia Torres-Sovero, our bird guide, were walking quietly down the trail from Posada Amazonas in Madre de Dios, Peru. Mother of God—what a name, so perfect for this apparently limitless Amazonian forest. The reserve I was in was managed by Ese Eja members of the Infierno Native Community. It is in an area I had dreamed of visiting, long before I gained a delightful Peruvian son-in-law and the chance to visit Arequipa to meet his family.

The walk was quiet at first. Had we begun before the birds had left their roosts? We walked on. Suddenly we heard *chips* and saw leaves stirring. I saw what Claudia identified as a Bluish-slate Antshrike. A Gray Antwren followed. Then I saw a Plain-winged Antshrike and an Olive-backed Foliage-gleaner. They are all somewhat dull birds unfamiliar to my North American eyes, but they are unmistakably different species. The female Bluish-slate Antshrike has a robin-like rusty belly, and the Olive-backed Foliage-gleaner a very long tail and a patch of white under its throat.

We stopped and watched the birds forage, each on its own as it hunted understory insects in the still-dim forest. Though each worked independently, together they clearly formed a group and maintained contact with one another through their soft voices. Within a few minutes, they had moved out of view, and the forest was quiet again.

To my untrained eye, the flock I saw did not seem to have any structure. But different birds play different roles. The first European to write about forest mixed-species flocks was Henry Walter Bates, who recounted his famous 1848 trip to the Amazon with Alfred Russel Wallace, co-discoverer of the theory of evolution by natural selection. Bates is perhaps most known for lending his name to Batesian mimicry, a phenomenon by which palatable species like the viceroy butterfly trick predators into avoiding them by looking like bad-tasting or toxic species like the monarch butterfly.

I downloaded Bates's classic book *The Naturalist on the River Amazons*, looking for the full passage that so many ornithologists quote as the first writing on mixed-species flocks.[1] I found it in the chapter entitled "Animals of the Neighbourhood of Ega" (now called Tefé). It is worth quoting at length here. I have omitted the outdated, racist language, both unnecessary and harmful.

> *Whilst hunting along the narrow pathways that are made through the forest in the neighbourhood of houses and villages, one may pass several days without seeing many birds; but now and then the surrounding bushes and trees appear suddenly to swarm with them. There are scores, probably hundreds of birds, all moving about with the greatest activity—woodpeckers and Dendro-*

> *colaptidae (from species no larger than a sparrow to others the size of a crow) running up the tree trunks; tanagers, ant-thrushes, humming-birds, fly-catchers, and barbets flitting about the leaves and lower branches. The bustling crowd loses no time, and although moving in concert, each bird is occupied, on its own account, in searching bark or leaf or twig; the barbets visiting every clayey nest of termites on the trees which lie in the line of march. In a few minutes the host is gone, and the forest path remains deserted and silent as before. I became, in course of time, so accustomed to this habit of birds in the woods near Ega, that I could generally find the flock of associated marauders whenever I wanted it. There appeared to be only one of these flocks in each small district; and, as it traversed chiefly a limited tract of woods of second growth, I used to try different paths until I came up with it.*

Bates goes on to discuss what the Native Americans know about these flocks and the birds in them:

> *The Indians have noticed these miscellaneous hunting parties of birds. . . . They have . . . a theory . . . to the effect that the onward moving bands are led by a little grey bird, called the Papá-Uirá, which fascinates all the rest, and leads them a weary dance through the thickets. . . . I could never get a sight of this famous little bird in the forest. I once employed Indians to obtain specimens for me; but, after the same man (who was a noted woodsman) brought me, at different times, three distinct species of birds as the Papá-uirá, I gave up the story as a piece of humbug.*

Bates did not consider the alternative—that different species of birds played the role of papá-uirá in different flocks, and that people who had lived their whole lives in the forest likely knew what they were talking about. Maybe Bates misunderstood what the local people were telling him.

Bates clearly viewed mixed-species flocks as a biological phenomenon worthy of comment, but these flocks might be easy to misinterpret as random gatherings and not organized and systematic groups. It seems that the mixed-species flocks of the neotropics are especially stable, which makes them particularly visible. Once mixed-species flocks were recognized as an entity, they were observed and studied all over the world. As a baseline, let's delve into some of the early studies of these flocks in the New World.

One classic study is Martin Moynihan's work on flocks on Barro Colorado Island.[2] The island is well-known to biologists. Originally formed by damming done during the construction of the Panama Canal, Barro Colorado Island has since become a nature reserve and the site of the famous Smithsonian Tropical Research Institute. Thousands of biologists have taken advantage of its diverse and pristine ecosystem to study many aspects of tropical biology. (My first visit was with a tropical ecology class in the summer of 1974, when I was a graduate student just starting at the University of Texas at Austin.)

For his monograph, Moynihan studied mixed-species flocks of Plain-colored Tanagers, Palm Tanagers, Blue Tanagers, Golden-masked Tanagers, and Green Honeycreepers. His tools were simple: time, attention, and pencil and paper.

From Moynihan's descriptions, it seems that on Barro Colorado Island, the Plain-colored Tanager took the role of the papá-uirá described by Bates's Native informants. Perhaps other birds preferred to follow Plain-colored Tanagers because of their noisy alarm calls; birds near others with strong calls can spend more of their time foraging and less time watching for predators. Indeed, Plain-colored Tanagers were warning their own families while other species eavesdropped. Though this is the classic study, I wonder how much the isolation of the island by the lake impacted the mixed-species flocks. As I describe toward the end of this chapter, the slightest forest perturbation can disrupt mixed-species flocks.

Fifteen years after Moynihan, Judy Gradwohl and Russell Greenberg revisited mixed-species flocks on Barro Colorado Island. Their focus was on flocks in the understory. These included Checker-throated Antwrens, Dot-winged Antwrens, and Slaty Antshrikes. Though their birds were less accessible than Moynihan's, Gradwohl and Greenberg had an important advantage: They put brightly colored bands on the birds, along with metal bands that identified each by number. With these markers, the researchers would follow individuals over six years of study, ultimately spending a remarkable 3,200 hours per person on the project.

Like Bates's flocks in the Amazon, Gradwohl and Greenberg's mixed flocks had very stable territories that did not vary in size or location over the six years of their study. In the understory, the antwrens were the papá-uirá birds whose danger warnings were followed by others,[3] though these researchers (having not extended the concept of papá-uirá as I have) talk of *nuclear* birds, those essential to a mixed-species flock. The nuclear birds tend to include both the papá-uirá and their consistent followers.

A similar study was conducted in another iconic field station called Cocha Cashu in Amazonian Peru.[4] At this site, six species maintained joint territories in the forest understory: Dusky-throated Antshrikes, Bluish-slate Antshrikes, White-flanked Antwrens, Long-winged Antwrens, Gray Antwrens, and Rufous-tailed Foliage-gleaners. Other birds sometimes joined these mixed-species flocks, too. Here, the antshrikes were the papá-uirá—dull-colored, active, noisy, and the first in the flock to move to new positions. They gave the loudest warning calls when they perceived nearby predators (usually one of Cocha Cashu's six bird-eating hawk species).

As in the other studies, Charles Munn and John Terborgh's flocks were stable, forming early in the morning, lasting until late afternoon, and remaining together all year, whether it was the breeding season or not. The core members of a given flock shared territories completely. Interestingly, each flock contained only a single pair of each species, and all species participated in territory defense.

There are mixed-species flocks elsewhere in the world. John Miall Winterbottom studied mixed-species flocks in Zambia, which at the time was called Northern Rhodesia.[5] Winterbottom began with a list of 107 groups he identified and noted what species were in them. He divided the birds into nuclear species (among which the most common were Green-capped Eremomela, Violet-backed Flycatcher, Damaraland White-eye, Black Tit, and Southern Helmet-shrike; drongos were also important), other regular species, and species only occasionally found in the flocks. Winterbottom suggested that the gregarious species are the nuclei, particularly the Southern Helmet-shrike (now merged with the White-crested Helmetshrike). From the way Winterbottom described them, the nuclear species filled the role of papá-uirá.

Though I find the neotropical mixed-species flocks to be the most cohesive, mixed-species flocks are also common in temperate regions. As I stand at my desk in Leland, Michigan, and look out at the feeders, I see that they are not always busy. Then one, two, even three Black-capped Chickadees fly in, choose a sunflower seed, and retreat to pound it out of its shell in the nearby spruce or deep in the branches of a willow hanging behind the feeder. A Tufted Titmouse might join in. White-breasted Nuthatches and Downy Woodpeckers also arrive, though not as often; they are more likely to cower in the willow branches than the more brazen chickadees and titmice. This is to say, on any given afternoon, I can watch as my feeders are visited by a mixed-species flock.

Douglass Morse conducted a thorough study of North American mixed-species flocks from 1962 to 1968.[6] Morse's birds are ones I know well. Each species on his list jumps off the page for me, each doing its little dance and then fading as the next comes forward. Morse worked primarily in the mature deciduous forest near Louisiana State University, in an understory that is flooded for about half the fall and winter. He observed twenty flocks, nearly all of which contained Carolina Chickadees and Tufted Titmice. I can just imagine them flying through the trees, going from one branch to another, hanging and probing and chattering so much of the time. In my experience, the Tufted Titmice would be a little higher up, but Morse did not report that.

The next most common birds were Downy Woodpeckers, Ruby-crowned Kinglets, and Myrtle Warblers. The kinglets, impossibly small, are often hard to see, but they, too, are talking much of the time, and that is usually how I notice them. It seems almost odd that a woodpecker would be in with these songbirds, but Downy Woodpeckers nearly always are.

Farther north, Morse studied mixed-species flocks in the forests of Maine and Maryland in similar ways, recording and tabulating flock members. The late-summer flocks he recorded in Maine had twenty-three species in the mixed coniferous and deciduous forest, thirteen of them warblers. Over half of those flocks contained Black-capped Chickadees, Red-breasted Nuthatches, Golden-crowned Kinglets, Black-and-white Warblers, Magnolia Warblers, and Black-throated Green Warblers.

Compare those generous numbers for a North American summer to the harsher circumstances of a Maine winter, when so many species have migrated away. There the twenty-four winter flocks Morse studied in the mixed habitat had only eight species. Nearly all the flocks had Black-capped Chickadees, and three-quarters had Golden-crowned Kinglets.

I love watching winter flocks when there is snow on the ground and not a leaf to be seen on a tree, only the needles of pines and spruce knifing the icy air. There is not much food in these cold woods, but the birds manage. Just thinking about them makes me wonder where I've tucked away my cross-country skis.

In his exceptional study, Morse showed that the birds used different parts of the habitat when they were in flocks than when they were actually alone, presumably because flocks are better at detecting predators (in flocks, birds can be more daring about where to forage). For example, Brown-headed Nuthatches foraged farther out in the twigs when Pine Warblers were present. Golden-crowned Kinglets foraged farther from the tree trunk when there were more Tufted Titmice in the flock.

Mixed-species flocks may reduce predation risk, but that reduction in vulnerability does not necessarily apply equally to all the flock members. Since birds are of different species, and they forage in different ways, they are less likely to compete for food. A forager that flies out to hunt insects, for example, may be more likely to detect predators than a bird with its head buried in bark or riffling through leaf litter, making the latter kind of bird likely to benefit from the alarm calls of the former. But the birds that give alarm calls may also stand to benefit if they can steal prey from others.

Hari Sridhar and his team looked at where birds positioned themselves in mixed-species flocks in Anshi National Park in the Western Ghats mountains of southwestern India.[7] They reasoned that birds that were very close to others might be positioning themselves to snatch a tasty bite when the others uncovered it. Such proximity was not needed in order to benefit from alarm calls. They watched for close associations among species by doing what they called "focal neighborhood scans" to see who was closest to whom at a snapshot of time. They did more than two thousand of these scans.

The birds that moved close to birds of other species were the Greater Racket-tailed Drongo, Ashy Drongo, Bronzed Drongo, Black-naped Monarch, and Indian Paradise-flycatcher. These five species were all what the researchers called "sallyers," because, like flycatchers, they sallied forth to catch prey rather than beaking it out from crevices or curled leaves. Interestingly, these birds foraged particularly close to others so that they might catch a morsel missed by the bird that gleaned it out of a leaf or from the soil. I guess it is no surprise that the favored associates of the sallyers were birds that foraged in other ways and might scare out insects the sallyers could catch. For example, Ashy Drongos associated with Orange Minivets, Black-headed Cuckooshrikes, and Eurasian Golden Orioles, while Paradise-flycatchers followed Western-crowned Warblers. These sallyers are extremely good at spotting predators, usually hawks, and then giving a loud alarm call to warn others. Some of these sallyers, such as the drongos, have, however, been known to give false alarm calls to get other birds to drop their prey. Then the sallyers snatch it up. These might be called the papá-uirás because their alarm calls attract other birds for their protection. But there is more to the story.

Sridhar told me that the nuclear birds are not always easy to identify but seem to be birds that give alarm calls readily, like many of the drongos aforementioned. But they are not the only ones that could be considered important to group cohesion. There are other birds that vocalize a lot and send out alarms even though they are not sallyers.

Sridhar said, "My hunch is the Brown-cheeked Fulvetta is the most important 'nuclear species,' but other species like the Scarlet Minivet in the canopy and Dark-fronted Babblers in the understory are also 'nuclear' often." He went on to hypothesize that the nuclear species were likely to be those that were gregarious within species, unlike those species in the neotropics.

I was fortunate to get a glimpse of some of the birds in these mixed-species flocks on a bird-watching trip in southern India after a meeting in Bengaluru at the National Centre for Biological Sciences in October 2016. My husband and I flew to Kochi in Kerala and traveled to the bird retreat of the late renowned bird guide Eldhose on a small rubber plantation. Eldhose told me that he had birded for seventeen years without binoculars. When David Attenborough came to his reserve to film the Malabar Pied Hornbill and realized his host did not have binoculars, he gave him a pair.

When I visited Kerala, I did not yet know about its mixed-species flocks, but looking back on my eBird notes, I can see I spotted some important birds for these types of flocks. For instance, I saw a single Brown-cheeked Fulvetta on October 25, 2016, on the Thattekad to Urulanthanni trail. My notes read, "We kept circling this one hotspot, a clearing in the forest along the road caused by a recently fallen tree, but also walked some along the road and along a trail to where a tribal group, the Muthuvans, were." We also saw Orange Minivets four times from October 23 to October 26, 2016, in the same general area and even on the same walk as the Brown-cheeked Fulvetta. I saw a Scarlet Minivet on a different trip to India, far north of the other birds that I saw in Kerala. This one was in Assam, in Kaziranga National Park, which I visited with Krushnamegh Kunte and Deepa Agashe before another Bengaluru meeting. I saw all five of the prominent sallyers identified by Sridhar and his colleagues, though I didn't know at the time to pay attention to their mixed-species flocks.

To look at the role of predation in maintaining mixed-species flocks, we can turn to a field station in central Finland. At the Konnevesi field station at the University of Jyväskylä, Jukka Suhonen took advantage of a special feature of the local predators to measure their success against multispecies flocks. At Konnevesi, Crested Tits and Willow Tits formed flocks that sometimes also included Goldcrests, Eurasian Treecreepers, Great Tits, and Coal Tits.[8] The flocks' primary predators were Pygmy Owls, which cached their victims in hollow trees or in wooden nest boxes nailed to trees by researchers. Researchers could count and weigh the prey in nest boxes to measure how much and which prey the owls were collecting. In the winter of 1988, the owls cached mostly voles and other small mammals, but the following year, the vole population crashed and the owls switched their focus to birds. In just a year, birds went from making up only 2 percent of the owls' diet all the way up to 60 percent. Nearly all the killed birds were the most common species in mixed-species flocks—Crested and Willow Tits.

The difference between these two years gave Suhonen a chance to compare bird behavior in a high predation year and a low predation year, a kind of natural experiment. In the high predation year, he saw birds move closer to the dense branches near the trunk of the tree or lower down on the tree—places safer from predation. In these flocks, Crested Tits were dominant to Willow Tits, and males dominant to females, so the most vulnerable birds were female Willow Tits, particularly the young ones who had hatched the previous summer. These birds got the worst places to forage. Their recourse was to forage lower, where there was less food; only occasionally did they venture to the vulnerable outer twigs. I was surprised that the owls were hunting during the day while the birds were still foraging, but I suppose it makes sense during the very long, dark Finnish winter.

Another study tested the predation hypothesis for mixed-species flocks experimentally in North America. It asked what would change about bird behavior if the papá-uirá were not there. In the winter of 1997, Andrew Dolby and Thomas Grubb Jr. used feeders to understand how important Tufted Titmice were to the foraging of another, less-dominant mixed-species flock member, the White-breasted Nuthatch.[9] They reasoned that the nuthatches would be less brave at visiting feeders when the titmice were absent. They conducted the study on ten woodlots surrounded by croplands in Union County, Ohio. At each site, Dolby and Grubb put feeders out with sunflower seeds or suet either 25 feet (7.6 meters) or 52 feet (15.8 meters) from the forest edge.

During the experiment, the researchers caught and temporarily removed the titmice from some plots and not others, then compared the sites. If Tufted Titmice warned others of predators, then the nuthatches might venture farther from the safety of the forest and into the meadows. To figure out whether they were right, the team looked at how quickly the White-breasted Nuthatches went to the feeders. When the feeders were only 25 feet from the forest edge, it did not matter to the nuthatches whether titmice were present. The nuthatches flew readily to feed. But when the feeders were 52 feet away, the nuthatches delayed going to the feeders if the Tufted Titmice had been removed. The experiment showed that the White-breasted Nuthatches were less fearful of undetected predators when the Tufted Titmice were around.

The Dolby and Grubb experiment explored fear and caution. Another kind of experiment looked at the responses of birds when there was an actual predator present, even just a stuffed Sharp-shinned Hawk being tugged on a wire. Kimberly Sullivan explored behavior in mixed-species flocks in the Great Swamp National Wildlife Refuge in New Jersey. There, the winter flocks contained mostly

Black-capped Chickadees, Tufted Titmice, Downy Woodpeckers, Hairy Woodpeckers, and White-breasted Nuthatches. The nuclear papá-uirá species were Tufted Titmice and Black-capped Chickadees, who gave incessant contact calls.[10]

Sullivan focused on Downy Woodpeckers, one of my favorites. The woodpeckers were either marked with colored plastic or identified by natural variation in their feather patterns. Sullivan figured that if the birds were in a flock, they could rely on others to warn of predators. She measured vigilance in Downy Woodpeckers by looking for head cocking. A bird that paused and cocked its head was being vigilant, listening, taking time away from feeding.

Downy Woodpeckers cocked their heads and looked around less in groups than they did when they were alone. This meant they had more time for feeding. The predators they were trying to avoid were Cooper's Hawks, Sharp-shinned Hawks, and American Kestrels, all birds Sullivan saw in these forests. Sullivan then investigated whether the woodpeckers were using the alarm calls of chickadees and titmice by introducing models of the hawks when the Downy Woodpeckers were alone or with flocks. She used a stuffed Sharp-shinned Hawk on a pulley suspended between two trees on a wire, launching the hawk and dragging it right back to the trees while she watched from a blind. She found that Downy Woodpeckers froze, then looked around and stopped foraging either when a predator (natural or stuffed) was sighted or when the chickadees or titmice gave an alarm call. The woodpeckers ignored the calls of sparrows in the underbrush.

Black-capped Chickadees gave alarm calls in all encounters with living raptors and in nearly all encounters with the stuffed model. Tufted Titmice gave alarm calls in encounters with live raptors and in half of the encounters with the stuffed hawk. So the Downy Woodpeckers had plenty of alarm calls to respond to, and they did not utter their own unless they had a mate in the group.

Sullivan recorded the woodpeckers' responses, noting that when the Downy Woodpeckers could hear contact calls from flock members, they reduced their vigilance. Contact calls are a kind of low chatter or chip that lets others know where a bird is. They are very different from the loud alarm calls that cause Downy Woodpeckers to freeze in place while the chickadees and titmice immediately dive into low bushes.

Ari Martínez and his team conducted experiments on the importance of what they called the "sentinel" (again, what we call the papá-uirá) in Amazonian forests.[11] They worked at the Pantiacolla Field Station in southeastern Peru on the north bank of the Alta Madre de Dios River. It is a wild place, reached by driving down the Cusco Road and then up the river by boat to a place where one lives in a tent and washes in the river, nary an Internet signal to be had. The team spent a lot of time there, identifying about thirty flocks and banding 70 percent of the birds in them. The flock membership was stable. The Martínez team engaged trained raptors from an expert falconer as their experimental predators (these birds had been trained to eat only from a person's hand, so they could fly over a flock without actually endangering it). Among them were two small raptors, an Aplomado Falcon and Bicolored Hawk, and a large one, a Harris's Hawk. The Bicolored Hawk is the local bird most likely to prey on birds, while the other two stood in for local raptors they might have experienced.

For the experiments, the team used sixteen mixed-species flocks, half led by Bluish-slate Antshrikes and half by Dusky-throated Antshrikes. These two papá-uirás did not occur in the same flock, and the Bluish-slate Antshrike was in younger patches of forest. Each flock experienced each raptor three times—once at close range, once at medium distance, and once farther away—though only once per day, and none of the flock members were harmed. Martínez and colleagues found that the Bluish-slate Antshrike was more likely to respond to the raptor with an alarm call than was the Dusky-throated Antshrike, which alarmed more to the smaller raptors, the ones more likely to eat small birds. Of course, the flock members responded to the alarm calls and hid.

The Martínez team did another experiment in which they removed the papá-uirá, in this case Dusky-throated Antshrikes, and then looked at the response of the flocks.[12] In this test, they used eight flocks, four control and four experimental (with the antshrikes removed). They focused on the responses of White-flanked Antwrens; these flocks were comparatively easy to study because they roosted together and left the roost every morning to forage. To make the control and experimental flocks as similar as possible, they captured all the Dusky-throated Antshrikes, then immediately let the birds in the control flocks go. (The others were kept in small aviaries, where they were fed grasshoppers for the three days of the experiment, then released back into their territories.) Sometimes, antshrikes that did not have their own territory tried to join the flocks without antshrikes. The researchers managed to deter them by playing territorial antshrike songs. The researchers found there were big differences during the time the flocks did not have their papá-uirás. They foraged in areas with deeper vegetative cover, and they were less likely to forage with the flock at all. No one would be able to measure it, but I bet the birds were very glad to get their papá-uirás back. Taken together, these two experimental studies provide strong support for the predation hypothesis for mixed-species flocks.

Since mixed-species flocks are groups of multiple species of birds, they may be more vulnerable to habitat loss, climate change, and other disturbances, just as larger animals are often more vulnerable than smaller ones. Studies that document habitat loss can be challenging, however, because they need to demonstrate conditions before and after. One of the best was undertaken by Mathilde Jullien and Jean-Marc Thiollay, who followed eleven mixed-species flocks over three years in a primary forest in French Guiana, 60 miles (97 kilometers) south of Cayenne, at a French Research Station called Nouragues.[13] They watched flocks from dawn to dusk, noting where the flock was every ten minutes. They usually saw seven species (Rufous-rumped Foliage-gleaners, Chestnut-rumped Woodcreepers, Dusky-throated Antshrikes, Cinereous Antshrikes, Long-billed Antwrens, Gray Antwrens, and Brown-bellied Stipplethroats) that were present in all flocks they studied and *only* occurred in mixed-species flocks. The antshrikes—particularly the Cinereous Antshrikes—were the papá-uirá that the others followed. They uttered warning calls against predators, allowing the other species to go about their leaf-gleaning and insect-searching business in peace, but they also sometimes used the confusion that the calls created to steal other species' food. Each flock contained just a pair of each species, sometimes along with their young.

The territories for mixed-species flocks during the three

years of Thiollay and Jullien's study were so constant that in 2011 another research team, Ari Martínez and Juan Gomez, decided to visit to see how the flocks' territories had changed eighteen years later.[14] They did not expect to see the same color-banded individuals but thought that the group territories might have persisted over nearly two decades if birds in a given mixed-species flock were replaced slowly. Indeed, this was the case. The layout of the territories was the same as before. A territorial border in 1993 was still a territorial border in 2011, no doubt maintained over the long period because the flock members did not all perish at once. The Martínez team found that fidelity of territories especially in understory flocks, but also in overstory flocks at Cocha Cashu Biological Station in Peru, where they compared territories between 1982 and 2018.[15]

But human interference can have sudden dramatic impacts, something that Jean-Marc Thiollay studied in a forest not far from the field station where he knew the flocks so well. The disturbance—selective logging—was supposed to be a mild one,[16] and he was optimistic about limited harms since natural gaps such as those created by tree falls, rocky outcrops, or riverbanks were known to be important for the diversity of the forest (they create opportunities for light to reach plants that cannot compete with canopy trees). So it was with some hope that Thiollay looked at birds in forests one to two and eight to twelve years after selective logging. Since he waited at least a year after the selective logging, no differences could be attributed to the machinery of active sites.

Sadly, Thiollay found that members of mixed-species flocks were among the most damaged bird populations. After logging, only half of the species were left. Thiollay saw decreases in foliage-gleaners, antshrikes, and antwrens in particular. Overall, mixed-species flocks were rare in selectively logged forests, suggesting that their foraging strategies were dependent on the dim light and stable microclimates of mature forests. Thiollay also visited sites that had been logged forty, eighty, and one hundred years ago and found they had not recovered to the kind of vegetation structure necessary for deep forest antshrike-led mixed-species flocks. The damage caused even by selective logging is profound and long-lasting.

In another approach to understanding the conservation of the members of mixed-species flocks, a team led by Fasheng Zou collected all the studies they could find on mixed-species flocks.[17] They concentrated on studies that identified the species and how many individuals were involved in mixed-species flocks. They found 201 published studies with these data from all continents except Antarctica.

Upon review, Zou and his team found that 2,049 species of birds were reported to participate in mixed-species flocks—some 20 percent of the world's 10,680 species of known birds. Among these were 1,705 songbirds and 174 woodpeckers, as well as 158 species on the International Union for the Conservation of Nature's Red List of Threatened Species. The Red List is a comprehensive assessment of the extinction risk of a species, taking into account things like range, population size, habitat, and threats. Of the mixed-flock species on the Red List, 2 were listed as critically endangered, 14 were endangered, 37 were vulnerable, and 105 were near threatened. These numbers may seem high, but it turns out that birds in mixed-species flocks are actually less likely than birds abstaining from mixed-flock foraging to appear in one of the categories of concern.

Thirty-one of the studies Zou and his team found specifically looked at the impact of human activities on mixed-species flocks. The studies examined the number of species per flock and the number of individuals per flock and reviewed how they differed according to human disturbance. It turned out, no surprise, that locations impacted by humans had seen a significant decline in the number of species per flock or in the number of individuals per flock. Overall, species in mixed flocks appeared to be more sensitive to human disturbance than other birds. The most disturbed sites had only three-quarters of the species as the least disturbed sites. The decline in species and numbers in the disturbed sites is most likely due to impacts on the 190 nuclear species the researchers identified. It is disturbing that so many birds are threatened by humans.

William Marthy and Damien Farine looked at one specific kind of human impact on mixed-species flocks: the songbird trade.[18] They compared mixed-species flock structures over twenty years in Sumatra, Indonesia. Of the roughly ninety species they looked at, forty-nine overlapped as both part of the bird trade and important members of mixed-species flocks. In Southeast Asia, birds are captured and sold as pets, for singing competitions, for religious purposes, and as food and medicine. Millions of birds are traded, and billions of dollars are involved in this desecration.

At the Way Canguk Research Station in Bukit Barisan Selatan National Park in Sumatra, Indonesia, Marthy and Farine surveyed the birds in 1997 and again in 2016 to see if flocks changed, if this was particularly true for traded species, and how the other birds fared when the traded species diminished. Though surrounded by degraded areas, the study area was generally undisturbed and in a large virgin forest—but right at the field station and in the area of the study, illegal trapping continued, to the immense frustration of station staff. Two important birds that usually occurred in mixed-species flocks were rare, the White-rumped Shama and the Grey-cheeked Bulbul.

The intensive analysis of the differences in flocks after so much bird harvesting showed that flocks contained fewer species in 2016 than they did in 1997. Mixed-species flocks were suffering because of the loss or reduction of traded species; increases in this harm will impact both traded species and others in the mixed-species flocks. As flocks become more diffuse, they function less effectively. Unfortunately, there is no obvious sign of improvement, and the illegal bird-trade business is very difficult to study.

One of the most important studies of the impact of humans on mixed-species flocks is part of the Biological Dynamics of Forest Fragments Project. It was launched in 1979 with Thomas Lovejoy as the lead researcher.[19] With support from the Brazilian National Institute of Amazonian Research, Lovejoy chose a study area 50 miles north of Manaus on the Amazon River. The original site was in an area planned for forest destruction (for cattle ranching), but the researchers received permission to preserve certain fragments of forest for study. First, the researchers documented the birds, mammals, insects, and plants in the untouched forest. Then the land was cleared, leaving intact isolated fragments of 250 acres, 25 acres, and 2.5 acres (100 hectares, 10 hectares, and 1 hectare).

In 2011, a team led by Karl Mokross visited the Biological Dynamics of Forest Fragments Project to study mixed-species bird flocks. At that time there were the original undamaged forest fragments, pastures that had once been burned, and areas that had been clear-cut but not burned. Since the 1990s, cattle were no longer in the area, and all the cut areas were regenerating. Mokross and his team followed twenty-one flocks across the various habitats.[20] They noticed that mixed-species flocks in the permanent forest and 250-acre forest fragments were very different from those in the smaller fragments or the secondary forest. (Secondary forests are those that grow back after the primary forest is cut down. They may look much like primary forests, but their trees are younger and not necessarily of the same kinds.) The mixed-species flocks in the untouched forest had richer and more complex interactions and were composed of more bird species. The disturbed-area flocks also had fewer nuclear birds.

Three species were in all habitats—Cinereous Antshrikes, Chestnut-rumped Woodcreepers, and White-flanked Antwrens—making up the core of mixed-species flocks. The Cinereous Antshrikes foraged by sallying forth and catching insects. They were the most likely to be in positions to see predators and give alarm calls. The Chestnut-rumped Woodcreepers, meanwhile, ran up and down trees, probing bark for insects, and the White-flanked Antwrens foraged by gleaning along leaves and branches, only sometimes sallying forth from a perch. With this in mind, I would say the Cinereous Antshrike is the papá-uirá, and the other two species benefit from its vigilance.

Besides these three ubiquitous species, the secondary forest lacked the birds of the primary forest. The primary forest has a rich group in the mixed-species flocks with about ten species common in the flocks and another two dozen often found.

The fragility of mixed-species flocks is no surprise to

experienced neotropical ornithologists who have discovered that these flocks avoid the open and will not readily cross even narrow roads. They disappear with selective logging and do not recover for ages. No wonder small forest fragments cannot maintain the rich form of sociality known as mixed-species flocks.

These days I see mixed-species flocks everywhere, and I love to watch as the birds busily forage for insects under the umbrella of a noisy leader who will sound an alarm should a predator come close. I think back to those long-ago days when Henry Walter Bates was first exploring the Amazon basin, before massive habitat destruction. I hope these beautiful flocks can thrive in what woodlands remain.

CHAPTER 4

Colonies

Safe Places near Food

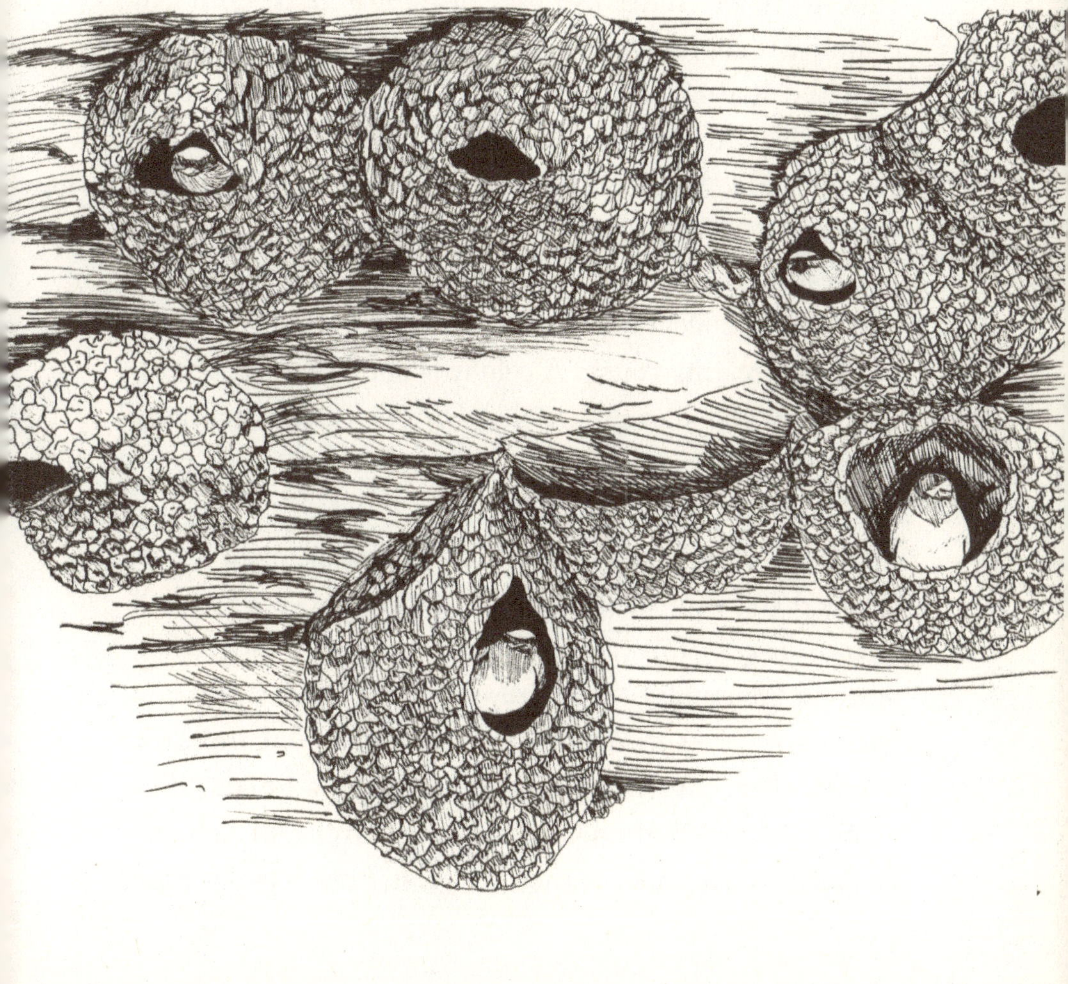

Standing on the shore of Lake Michigan and watching Bank Swallows fly in and out of nest holes in the steep wall of a sand dune, I feel that there is nowhere else I would rather be. I look across the lapping water to the Manitou Islands and wonder if Bank Swallow colonies are there, too. There are so many birds in these colonies, something is always happening, and my attention flits from one nest hole to another.

Colonies must be a successful nesting strategy, since a fifth of all bird species nest colonially, or in fixed locations where pairs of birds nest in close proximity. A colony, like that of the Pied Kingfishers I saw in Kaziranga National Park in Assam, India, can have just a few nests. Or it can comprise a copse of trees containing millions of birds, as with the Red-billed Quelea of southern Africa's arid regions. The birds may nest in sandbanks, like Pied Kingfishers and Bank Swallows do, or plaster their mud nests on cliffs and bridges, like Cliff Swallows, but a colony has two key features: First, nests are close together; second, each nest is independent of the other nests and has its own pair of parents rearing young. A colony is not an extended family. When the young have fledged, the parents generally leave the colony, though they may establish or join a non-breeding roost in the same area.

The colony I know best is on an island made up of dredged soil in a small reservoir at Smith Oaks in High Island, Texas. This site is famous for spring migration, wherein thousands of songbirds drop into the live oaks and hackberries after flying straight across the Gulf from the Yucatán Peninsula in Mexico. After flying all night, they arrive in the early afternoon; if they are strong enough and no north wind impedes their progress, they may keep on going. But if there is a north wind, many Scarlet Tanagers, Rose-breasted Grosbeaks, and warblers will rest, exhausted, in the trees of High Island. On those days, bird-watchers come to the Texas coast in droves—myself included. I try to visit toward the end of every April for a Houston Audubon Birdathon with my husband, sister, son Philip, and colleague and friend Cin-Ty Lee.

But it is the colonial birds of High Island that reliably astound. For them, this slice of land is not flyover country but valuable breeding ground. Step onto any of the half dozen or so observing platforms and behold an abundance of birds behaving as if no one is watching. The Houston Audubon Society keeps careful counts, and so we know, for instance, that on May 11, 2023, there were 744 Great Egrets, 571 Tricolored Herons, 552 Neotropic Cormorants, 530 Cattle Egrets, 500 Snowy Egrets, 160 Roseate Spoonbills, and 75 Little Blue Herons nesting in the rookery (an undercount, as the hundreds of youngsters are left out of the official tally). The birds are noisy, each with its own voice. I love the improbable Roseate Spoonbills, with their magenta, orange, and white coloring, their eponymous spoon-shaped bills, and their wise eyes. But I like them all, really—the Great Egrets and their nestlings, fighting one another for survival (as I described in my book *Slow Birding*), right alongside the prehistoric-looking Neotropic Cormorants, with their long necks and near porcine calls.

I could not bring the students in my bird behavior class at Rice University to see the wonders of the African plains or even the flamingo colonies in Yucatán. But I could drive them an hour and a half from Houston to High Island's Smith Oaks rookery, where they could observe colonial behavior in the wild.

You might think these birds are nesting too densely to get food easily, but they do not stay in the area for food. Instead, they fly off to the extensive marshes that surround High Island. To be clear, High Island is not an island at all. It's an underground salt dome, rising from the coastal marshes and offering migratory birds their first sighting of trees. The salt dome signals subterranean oil, so it is no surprise that the marshes around High Island are pocked by oil machinery, pumping away. Still, the marshes provide the fish that the rookery birds need to feed their young.

Why do the rookery birds nest so tightly together, one nest often just inches from the next? Protection from predation is the most likely answer. The rookery's location in an artificial lake provides protection from land predators, though opportunistic alligators lie in wait for birds that fall from their nests into the waters that surround the island. Birds in the center of the colonies are better protected than those at the edges, so as the birds arrive in the spring, they build their nests close together. The oldest nests thus end up being the most central. Joanna Burger reviewed this topic and found evidence of higher breeding success for central nests in Yellow-headed Blackbirds, Red-winged Blackbirds, Cattle Egrets, and swallows.[1] Central nests are not only the last to be encountered; they are also able to attract the largest mobs against predators.

Richard D. Alexander, my undergraduate mentor at the University of Michigan, wrote a landmark paper about why animals live in colonies. It was published in 1974, just as I graduated.[2] Dick, as I eventually dared to call him, met with me weekly for two years, guiding me through a series of readings that ultimately informed my undergraduate honors thesis on play behavior in non-primate animals.

Dick also generously shared his thinking, in conversation and on paper, from the first germ of an idea through many rough drafts. (Anyone of a certain age will remember the off-purple color and sweet-in-a-chemical-way smell of such mimeographed pages.)

Shortly after Dick's paper came Richard Dawkins's 1976 book *The Selfish Gene.*[3] Along with altruism within families (a topic for the second half of this book), *The Selfish Gene* dissects how natural selection on social organisms works and how conflict and cooperation dance together. Those folks unlucky enough to be outside Dick Alexander's attention were now, with Dawkins, lucky enough to learn the logic that underlies social behavior.

Indeed, in the 1970s biologists were trying to understand social behavior from new evolutionary angles that took the self-centered nature of behavioral evolution into account. Dick Alexander's big idea was that group living, common as it is, has costs for individuals because it increases intraspecies competition for resources and the transmission of diseases and parasites. That resource competition, by the way, is not only over food or nesting material in birds; it is also about mating and reproducing. Neighbors might mate with others' mates, dump eggs in others' nests, or even destroy nests altogether—selfish behaviors that benefit them but come at a cost to others. These issues were known to be factors for birds in flocks and roosts, but Dick thought they might be more serious for birds nesting in colonies, whose stationary nests put them in fixed close contact. Logically, he saw that colonial nesting would not evolve and persist if it didn't offer compensatory benefits.

In 1972 and 1973—in the middle of my undergraduate years at Michigan—Dick's second-year graduate students

Paul Sherman and John Hoogland decided to find evidence for the costs and benefits of coloniality.[4] They chose the Bank Swallows that commonly nested in the sandy walls of inactive gravel pits near Ann Arbor, Michigan. Once, these open-pit mines produced many tons of sand and aggregate to fuel a construction boom in southeastern Michigan. When left inactive, however, they served other tenants and provided many hours of fun for children like me. Willing to defy parents' strict rules against the dangers of drowning and cave-ins, my friends and I fished for bluegills and swam in the cold, spring-fed water that filled the gravel pits near my childhood home in East Lansing, an hour northwest of Ann Arbor. Today, I can still walk along the leveled shores of those gravel pits, though homes now line one side, and the car that had rusted in the water was fished out long ago.

The Bank Swallow colonies Hoogland and Sherman studied ranged from 2 to 451 nests. They noted that the swallows dug burrows and lined the chambers with grass, rootlets, straw, and small twigs, often stolen from the nearby burrows of other swallows. But these thefts would not compare with the thefts they'd observe when the birds began to lay eggs and further insulated their nests with feathers. Hoogland and Sherman describe the fights that broke out as birds sighted a colony member flying in with a prized feather in its beak. As many as a hundred other birds would swarm, and the feather could change beaks as many as seven times before getting tucked into a burrow. No bird feathered its nest without a fight. And just who benefited from the cushiony feathers? The young in the burrow did.

Besides competition for nest locations and materials, there is another very serious cost to group nesting, as I mentioned at the very beginning: parasites and diseases. The closer together the Bank Swallows are, the more likely they are to catch diseases from one another and to transmit parasites. (In these pandemic days, this is a lesson we humans have come to know all too well.)

Most pathogens and parasites are hard to count because they are tiny or inside the body, or both. But ectoparasites—those on the outside of the body—are an exception, and they have been widely studied. The most common ectoparasite that Hoogland and Sherman observed among the Bank Swallows was a bird flea that waited at the entrances of active burrows. To see if larger colonies had more fleas, as one might predict, Hoogland and Sherman counted fleas at the entrances of 191 burrows in 22 colonies of different sizes on the tenth day of the burrow's existence. As expected, there were more fleas per burrow in larger colonies. These fleas carry diseases but themselves are unlikely to be harmful, as discussed later in this chapter.

At this point, when everything indicates the social and disease costs of living in colonies, you might be wondering why on earth Bank Swallows do it? And why in such large colonies? Given that lone burrows are rare and most Bank Swallows nest in colonies of more than a hundred burrows, the researchers wondered what advantages bigger colonies offered. It became another topic that our indefatigable pair went after, methodically testing Dick Alexander's ideas.

Out of more than three thousand active burrows, Hoogland and Sherman found only one Bank Swallow nest that was by itself. (They did not check it again, so we do not know if the birds of that lone burrow successfully raised young. My guess is that they did not, and that the parents of that nest were themselves very young.) They found that the nests were clumped even in sandbank faces. Furthermore, they discovered that when a new nest was added, it was sited close to existing nests, and subsequent nests spread out from those nuclei.

Next, Hoogland and Sherman looked for social foraging. If birds nested close to one another so as to follow others to feeding locations, then they would be expected to leave and return together. This hypothesis was not supported: The birds did not follow others when they left the colony. They argued that birds do not need to nest together to notice and join successful foragers once they were already away from the nest. If having other birds in the area to follow was important, then birds in larger colonies should be able to find more food. But instead, birds in larger colonies were worse at foraging, at least as measured by both the survival and weight of the nestlings.

Hoogland and Sherman looked at this in two ways. First, they took advantage of an unusual four-day period of cold, rainy weather right at the height of breeding season, something that would make insects harder for parents to find and provide to their young. Thinking that this might be very bad for the swallows, the researchers went out and checked ten of their colonies. Sadly, they found many babies dead at the bases of the colonies. There was no evidence of external harm. Hoogland and Sherman concluded that the babies had starved to death. There were more dead per nest beneath larger colonies than smaller ones. The cold weather meant flying insects flew less, and therefore Bank Swallows could not get the food they needed at this crucial time, right in the middle of the breeding season. Presumably the increased competition for food near larger colonies meant more babies died there. In addition, at ten days old, surviving young birds weighed less in larger colonies.

Again, it looks like advantages to group nesting may be hard to find in the midst of all these negative effects, but there is one potentially huge advantage. Could it make up for all these disadvantages?

That powerful possibility is that birds nest together to escape predation. One way Bank Swallows might do so is by banding together to attack and mob predators. Indeed, Hoogland and Sherman found that most of the adults in a colony quickly mobbed potential predators like Blue Jays, crows, or kestrels, flying in a "doughnut-shaped vortex" several birds thick. They found that birds would even mob predators at nests in the colony far away from their own. We know now that mobbing happens at all nesting stages. The mobbing birds hover close to the predator for a few seconds but do not touch it. Sometimes predators kill a mobbing bird, so the behavior carries risk.

Hoogland and Sherman made several predictions about predation. All were based on the observation that adults in different burrows were not related to one another, so any helpful interactions ultimately had to benefit the birds in each burrow. If mobbing was protective, it should be directed only against actual predators like long-tailed weasels, Common Crows, Belted Kingfishers, Blue Jays, Sparrow Hawks, and the like. Birds that were never observed attacking Bank Swallow young—like other species of swallows, Mourning Doves, and Song Sparrows—should not elicit mobbing behavior.

Seldom witnessing actual predation, Hoogland and Sherman tried another approach. They obtained a stuffed weasel and introduced it to colonies. They predicted that larger colonies would notice and mob the weasel sooner than smaller colonies would. And sure enough, larger colonies vocalized and mobbed the weasel sooner after its introduction, and they recruited more birds to the mobbing crowd.

Colony size is not the only variable that proved important for predation—or at least, avoidance of it. A burrow in the middle of a colony was less likely to be attacked than one on the edge of the colony, for instance, so the birds nesting in the center should have more success than those nesting on the edges. Again, the researchers used their stuffed weasel and found that fewer birds joined the mob when the weasel was near peripheral nests than when it was near the central ones. For example, in a colony of 172 birds, 61 mobbed a central weasel, while only 45 mobbed a peripheral weasel, on average. The central nests have more neighbors and so are more likely to attract birds to mob.

Not so long ago, John Hoogland and I reminisced about those wonderful days when we, and our understandings of social behavior, were young, and we were working with Alexander, a leader in the field, at the University of Michigan. John reminded me of Dick's rules for successful fieldwork, which I had not thought about in a long time but still have taped to the cabinet in the lab where I keep flashlights, butterfly nets, clipboards, and other field gear. It is a photocopy of something clearly typed on an old-fashioned typewriter. Dick entitled the list "How to Stay Optimistic While Trying to Conduct a Field Study." John remembered "Live in the field" and "Think, but let the animals do the leading." He also liked "Have someone around to complain to (and blame it on)." My favorites were "Do the easiest things first" and "Have several things going at once so you can afford failure." Each bit of advice was a gem that helped me through many hot, uncomfortable days and nights in the field—especially when I was working on wasps as a graduate student.

Though John went on to spend his career studying prairie dog coloniality, he encouraged his graduate student at Princeton University, Charles Brown, to study swallows. They became Charles's life project. Charles told me he still remembers when John first suggested that he study Cliff Swallows, as it changed his life. About a decade after the Bank Swallow study, Charles and his then wife, Mary Bomberger Brown, began their study of Cliff Swallows in southwestern Nebraska in a 93-by-31-mile (150-by-50-kilometer) area around Cedar Point Biological Station. It is a beautiful part of the country with many bluffs along the North Platte River; John Janovy's *Keith County Journal* makes the area come alive.[5]

The Browns' study has been going on since 1982.[6] There is an illustrated web page for their Cliff Swallow Project (cliffswallow.org), if you want to see for yourself. On it, Charles describes these birds, the same species as the famous swallows that nest at the Mission San Juan Capistrano in California (having been chased away from stores by merchants). Though few swallows have actually nested at the mission in the past quarter century, the city of San Juan Capistrano still holds an annual festival in honor of the birds every March 19.

I first grew to love these birds in Central Texas, but it turns out Cliff Swallows were originally from the canyons and cliffs of the West, inhabiting the Rocky Mountains as well as the Sierra Nevadas and Cascades. What brought them east was the construction of bridges, culverts, and buildings, which provided nest sites. Year after year, they still move eastward.[7]

Cliff Swallows nest in much larger colonies than the Bank Swallows studied by Hoogland and Sherman do.[8] Charles and Mary Brown have found colonies of up to 6,000 nests, with an average of 393 nests per colony. The nests are mud domes with entrances toward the bottom edge. Their surfaces are uneven, showing each beakful of mud that went into their construction, and they tend to be stacked atop one another, most commonly in two or three rows of nests. The colonies are often found in human-built structures, and this has allowed the Browns to conduct many comparisons between larger and smaller colonies.

The Browns found that larger colonies sometimes suffered when nests were built on top of one another to the extent that newly built nests blocked the entrances of existing nests. This situation, which did not happen in smaller colonies, could entomb the eggs or young inside. In fact, in two cases an incubating adult was trapped in the nest and the Browns felt obliged to break through the mud wall to rescue it. They noted that these blockages occurred in larger colonies with more than 16 nests per square meter.

Another hazard in the larger colonies was egg tossing—sometimes birds pushed out an egg from a neighbor's nest. This did not happen at all in colonies with only 10 nests. The Browns and their team saw eggs tossed from 479 nests, mostly a single egg but sometimes multiple. They were puzzled: Egg tossing has a clear cost to the victims but no clear gain to the perpetrators. Maybe birds were preparing their neighbors' nests for another common behavior: adding eggs to neighbors' nests, or egg dumping. If a bird tosses an egg from a neighbor's nest, it is less likely the victim will notice a new egg laid in the nest. But the two events were not strictly linked in time, so tossing still makes no sense.

Females laid eggs in nests that were not their own early in the laying sequence and did so more in larger colonies. As with egg tossing, the Brown team found, there was no egg dumping in colonies of ten or fewer nests. But above that threshold, the more nests in the colony, the more likely eggs were to be added to nests by neighbors. In colonies of a hundred or more nests, at least 10 percent of nests had eggs laid by another female. Surprisingly, a host accepted these parasite eggs even when they were laid as many as four days before she started laying herself—that is, when she ought to know they are not her own. The hosts seemed to have noticed the extra eggs because they themselves then laid fewer eggs in their own nests.

Cliff Swallows don't just lay eggs in neighbors' nests. Sometimes a bird will even take an egg in their bill and fly from their own nest over to a neighbor's nest, pushing it in to be cared for by someone else. The Browns and their team noted that this was more common in larger colonies of one hundred to one thousand nests, that all the egg dumpers had nests of their own, and that nearly all the nests with parasite eggs were within five nests of the perpetrator's own.

As we've discussed, nesting in colonies increases exposure to diseases and parasites. Of these, the ectoparasites are the most prevalent. The Browns found four kinds of ectoparasites in their Cliff Swallow colonies: swallow bugs (closely related to bedbugs), fleas, ticks, and chewing lice. The first two were the most common, with swallow bugs being the most harmful, and fleas doing no discernible damage. Ten-day-old nestlings with swallow bugs weighed as much as 22 percent less than those without, though the story changed with time, as discussed later. Swallow bugs also transmitted viruses, one of which caused a variant of encephalitis (the impact of these viruses could not be measured separately from the impact of the swallow bugs themselves). Furthermore, nestlings were less likely to survive to ten days in nests with more swallow bugs.

The ectoparasite problems were worse in larger colonies. The Browns counted all the swallow bugs and fleas on nestlings when they were ten days old and found that larger colonies had a lot more of both. There was an astonishing tenfold increase in swallow bug parasitism and a fivefold increase in flea parasitism in the largest versus the smallest colonies.

If we look at entombment, egg tossing, egg dumping, and parasites, it would seem that larger colonies have huge disadvantages. And yet they are so common. Are there compensating advantages? The Browns found a number. For instance, though competition for mates was more common in larger colonies, actual fights were fewer. Colonies of more than one hundred nests had one or two fights per nest per hour, while colonies with fewer than one hundred nests had three or four fights per nest per hour, on average. A well-placed beak stab could kill a rival, but these are very rare. In colonies of all sizes, fights for central nest sites were more intense early in the season. Clearly the Cliff Swallows perceived an advantage to the central nests, or they would have dispersed their nests more.

Another advantage of large colonies is that birds follow others to good foraging locations from larger Cliff Swallow flocks, something Hoogland and Sherman did not see with Bank Swallows. The way it worked was that Cliff Swallows flying about would see other birds from their colony foraging and would join them at what seemed to be good sites for insects. (Such a site might be behind a haying tractor; Charles Brown suggested that following tractors might be similar to following the bison that once roamed the prairies.) Since the swallows joined others at feeding sites with plenty of insects, this kind of foraging-following would not lead to the problems of information centers pointed out by Doug Mock, discussed in chapter 2. An important feature of this sort of foraging is that insects are abundant in the places discovered by Cliff Swallows. This abundance would explain the foraging flocks of as many as two thousand birds that the Browns observed. It would also explain the special calls that Cliff Swallows use when they have found a lot of insects, attracting other birds to the swarm.

The Brown team found that the larger the colony, the more food the parents brought back per foraging trip. Nestlings in large colonies received more food per hour than did those in smaller colonies.

Beyond the foraging advantages, there are a couple of indications that larger colonies are better at dealing with predators. Birds from larger colonies of more than one hundred nests built their nests in three to nine days, while birds from colonies of fewer than one hundred nests took fifteen to eighteen days to build their nests. Since the larger colonies were even farther, on average, from mud sources, the Browns mostly attributed this difference to the fact that the birds in smaller colonies spent more time watching for predators—they had fewer other birds to share vigilance duty. They also noticed that nests in larger colonies were more likely to share walls with neighboring nests, which would reduce building time.

The predators to watch out for were mostly birds;

American Kestrels, Barn Owls, Great Horned Owls, Black-billed Magpies, and Common Grackles all attacked and ate Cliff Swallows, catching them on the wing or as the swallows gathered mud to build their nests. The Cliff Swallows seemed most alarmed at American Kestrels and Black-billed Magpies, and they mounted a strong mobbing response to each. But the biggest threat was actually a single Common Grackle, Charles Brown told me. He said, "A grackle attacked mud-gathering birds, and over a two-day period, the same one (known by a missing wing feather) killed fifty adult Cliff Swallows, and eventually a total of about seventy before the carnage stopped. He just walked up to them and grabbed them. Took a lot of casualties before they started responding to him as a threat. Most of the banded birds among those killed were first-year birds. The grackle was only eating brains with so many kills!"

Bull snakes, however, were much more of a threat than any of these birds. They managed to crawl down the bridges and enter the nests, eating whatever they found. The Browns saw one bull snake stay in a colony for three days, eating the contents of fifty nests (about 150 eggs). Indeed, in the colony above my wasp nests in Texas, I once saw a snake winding through the swallow nests as if they had been built for its easy access.

Snake predation was rare enough that the Browns could not relate colony size to it directly. Instead, the team did an experiment with a model snake to see how readily the Cliff Swallows detected it and whether colony size mattered. It did. Colonies of around a thousand birds responded to the fake snake when it was as far as 200 feet (61 meters) away. Smaller colonies did not respond to the snake until it was just 60 feet (18 meters) away, and the smallest colonies (one to three nests) never responded. So larger colonies are better protected from predation.

Apparently, the advantages to feeding, nest building, and predation are significant enough to explain large colonies and to counterbalance the increased levels of reproductive competition and ectoparasites. However, there is more to the swallow bug story, something that only became apparent after thirty years.[9]

Charles explained what might have been the biggest shock of his career. I have talked about all the harm that swallow bugs do to nestling Cliff Swallows. What if I told you that the harm has now disappeared, even though there are just as many swallow bugs as ever? That is indeed the case. Charles pointed me to his *Ecological Monographs* paper and a pair of photographs. The first, from 1984, showed a pair of ten-day-old nestlings. One came from a nest that was infested with swallow bugs, the other from a nest that had been fumigated. The nestlings looked nothing alike. The bird from the parasitized nest was tiny and bare-skinned, with few developed feathers. The bird from the fumigated nest looked three times as large and had feathers sprouting out all over. These days, the image is famous: It appears in many textbooks as a classic illustration of the harm ectoparasites cause.

The second photo in the *Ecological Monographs* paper pictured two nestlings in 2015. Again, one was from a nest with swallow bugs and the other from a fumigated nest. But this time, the two birds were indistinguishable. What changed? Brown surmised that the difference boiled down to the birds' reactions to the bugs. Over thirty years, their immune system had apparently evolved to somehow tolerate swallow bugs. Brown reasons that there was a big increase in swallow bugs as the swallows became more abundant in human structures; over the decades of his study, he was able to witness this evolutionary change.

Some might think that thirty years is too short to see an evolutionary change—this genetic transformation of the swallows. But another change, clearly evolutionary, had also occurred entirely separate from that of the immune system. It had to do with the artificial habitat of the Cliff Swallow nests.

In the decades when the Browns drove from one colony to another along remote Nebraska roads, they always stopped if they saw a Cliff Swallow that had been hit by a car.[10] They prepared professional scientific samples, called skins, by pulling the bird's innards through a small opening and then stuffing the carcass with cotton or a similar material. Over the decades, the Browns picked up many such roadkills.

But they noticed that between 1983 and 2012, the roadkill swallows dwindled. The Browns found around twenty a year in 1984 but just four in 2012. If the population had declined, that would explain it, but that was not the case. The population had actually grown along the route they drove, from around six thousand birds to nearly twenty-five thousand birds. The numbers of scavengers that might have removed the dead swallows before the Browns drove past did not seem to increase, and there was no decline in traffic on the roads. Had something changed that made the swallows better at avoiding vehicles? the Browns wondered. Eureka! When the researchers compared the birds that were killed by vehicles to a random sample of birds from the population at large, it turned out that the birds killed by auto strikes had longer wings. There was a difference at the beginning of the study in 1984, and the difference got bigger over time.

The birds that the Browns picked up as roadkill were no doubt only a subsample of all the birds that were killed or injured. This means that the evolutionary force was probably even greater than the small numbers of roadkill would indicate. It was a force strong enough to change the average wing lengths of the entire population over the thirty years of the study: Wing length decreased from around 4.4 inches (11.1 centimeters) in 1984 to 4.2 inches (10.7 centimeters) by 2012. This may not seem like a lot, but it is a powerful change that impacts flight ability.

What might shorter wings do for a bird? After all, wing length has evolved over the millennia and was presumably optimal for swallow bodies at the start of the study. What changed was the birds' environment. Birds nesting under bridges will be in frequent contact with vehicles; the better they get at turning quickly, the longer they will live. Also, they tend to sit on roads, so being able to fly up right away is advantageous. Shorter wings, as it happens, facilitate both agility and flying up from the ground. Just as the immune trait of tolerating swallow bugs shifted, the morphological trait of wing length provided a stunning example of natural selection in action.

Bank Swallows and Cliff Swallows have allowed researchers to assess the benefits and costs of large versus small colonies. But to assess the costs and benefits of colonial nesting itself, we can turn to a bird that only sometimes nests in colonies. The Fieldfare, *Turdus pilaris*, common in Europe, is one such bird. Fieldfares share a genus with American Robins, *Turdus migratorius*. The two species resemble each other in their upright posture and in the worms they feed their young, though they are easily distinguished by their songs. To my ear, the American Robin has a slow, melodious song and a high-pitched alarm. The Fieldfare's song is harsher, lower, and more repetitive, as is its alarm call. I saw Fieldfares by the hundreds in the hedgerows around Oxford, England, where in autumn they seemed to flow in overlapping waves through the thick bushes, chattering constantly.

Fieldfares' open-cup nests make them especially vulnerable to losing their nestlings to predators. Sometimes Fieldfare parents go it alone and defend their chicks by hiding their nests from predators, but sometimes they do something far more bizarre.

As if earning their peculiar genus name, *Turdus*, Fieldfares respond to threats from flying predators (like crows and owls) by mobbing together, flying at their foes from behind, braking by backflapping, and then blasting the contents of their cloacas onto the predators at close range.[11] Effectively, the maneuver glues the predator's feathers together with feces. Birds so depend on clean feathers to fly, it makes me wonder why more birds do not use this form of defense.

As diverting as defensive defecation is, let's get back to what makes Fieldfares especially useful for understanding the costs and benefits of colonial nesting. Volker Haas studied Fieldfares in southern Germany to compare the strategies of colonial and solitary nesting.[12] Fieldfares began to nest in the spring as soon as the soil was warm enough for earthworms to come close to the surface but before the deciduous trees regrew their leaves from the winter. With the trees bare, the Fieldfare nests were impossible to conceal, and Haas found that most birds nested in colonies, defending their nests together. Only 8 percent of the 445 nests he watched in the early spring were isolated. And as expected, nests in colonies were more successful than single nests, and larger colonies were more successful than smaller ones.

When early-season nests were destroyed by predators or accidents, Fieldfares built new ones. If the trees had leafed out, nest concealment became easier. When Haas repeated his experiment late in the season, 20 percent of 239 nests were isolated rather than in colonies. Among later-season nests, nesting success in colonies and in isolated nests was the same—colonies were more visible to predators but better defended, while isolated nests were better hidden if less well defended.

Christer Wiklund found a similar pattern in Fieldfare colonies in Padjelanta National Park in northern Sweden, near the Norwegian border.[13] There, Fieldfares nested in colonies that included not only other Fieldfares but also Merlins, a type of falcon too small to prey on the Fieldfares.[14] Both species seemed to derive mutual protection from their shared main predator, the Hooded Crow. They even launched joint attacks. Fieldfares nested exclusively in birches, choosing areas that already had Merlin nests when possible (Merlins nest earlier than Fieldfares). The birds that nested in colonies fledged at a younger age and were more likely to survive than those that nested alone. Furthermore, those birds nesting in the centers of the colonies produced more eggs and fledged younger than those along the colony edges. Central nests had a stronger mobbing defense against predators, but also had older, more experienced birds. The presence of Merlins resulted in higher Fieldfare fledgling success. In Sweden, as in Germany, group nesting was advantageous, particularly to the birds in the center of the colony.

These studies make a strong case that Fieldfares nest in colonies in order to protect themselves from predators, but other birds may have different reasons for colonial nesting. Henry Horn studied colonies of Brewer's Blackbirds in the Potholes region of the Columbia National Wildlife Refuge in eastern Washington state.[15] It is a dry area in the rain shadow of the Cascade Mountains, and Horn surmised that Brewer's Blackbirds had not begun breeding there until 1951, when irrigation created new, wetter habitat. Horn found Brewer's Blackbird nests in sagebrush near cliffs and high trees, where males could perch watchfully, and near water, where insects emerged. The colonies covered a large enough area that edge birds did not see what others were doing as clearly as center birds did.

Predators took fewer nestlings from the center of colonies. The birds defended against predators by mobbing them. This worked best with flying bird predators; the Brewer's Blackbirds could not deter snakes or mammals.

Brewer's Blackbirds succeeded in colonies more because of what they ate than because of what ate them. They ate and fed to their young damselflies as they emerged from the lakes, so the water's edge was the best place to forage. In any emergence zone, there were more insects than any one bird could eat, so watching other birds' foraging behavior could be advantageous. Horn found that young in nests at the center of colonies gained more weight per day than young from more peripheral nests, perhaps because they could see more birds and could follow them to good damselfly hatch areas. This weight-gain difference made Horn think foraging is an important reason for these blackbirds' colonial nesting. The birds may not tell each other where to forage, but leaving from a central location makes it easier to spot where others are having success.

A mistake that trips up many biologists is assuming that every trait or behavior must have an adaptive explanation. Such explanations make good stories and, when supported by good studies like the ones I've described, are a lot of what make biology so fascinating. However, is it possible that some bird colonies may not have anything to do with nesting together at all?

Joel Sachs, Colin Hughes, Deborah Buitron, and Gary Nuechterlein studied Red-necked Grebes on chilly northern lakes.[16] This is a team I know particularly well, because Colin Hughes was my first graduate student at Rice University. He worked with me on social wasps, but birds were his first love, so when he became a faculty member at the University of North Dakota, he turned to birds, applying his keen observational skills and his wizardry in a DNA lab.

The Red-necked Grebe breeds in small, shallow northern lakes and other waters through much of the northern hemisphere, then winters mostly offshore in marine inlets and bays.[17] A detail about their family life that I find adorable is that as soon as chicks hatch, before their feathers are even dry, they climb onto the back of the incubating parent and snuggle down under the parent's feathers. Both mothers and fathers carry the young in this way. The parent then swims off into open water with about three chicks, sometimes abandoning the last few eggs. The other parent brings them food. Parents frequently shake the young off their backs to give them a chance to defecate and explore, but they always let them back up. Chicks return often to their feathery shelter until they are a couple weeks old, though parents still feed them until they are six to seven weeks old.

Red-necked Grebes are highly territorial and chase away any other birds from their nests. If they have nested in a small pond, they won't even let other Red-necked Grebes enter that pond. They attack other species, often by submarining, or attacking from under water. Even so, Red-necked Grebes can sometimes be found in colonies. Why would this aggressive and territorial bird jettison its aggression and nest with others?

Sachs and his team worked on Lake Osakis, a 6,270-acre (2,537-hectare) lake in central Minnesota. The strong winds, up to 30 miles (48 kilometers) per hour, make this a harsh habitat, and the lake can be iced over until mid-April. When spring finally arrives, grebes appear and quickly build floating nests, typically in the shallows along the shore or on islands made of mats of cattails that have broken off from the lakeshore.

Sachs and his team carefully observed Red-necked Grebe nests built on the lakeshore, on small cattail islands, and on large cattail islands, studying the fate of the nestlings in each habitat. It turns out the birds saw much more success on cattail islands, especially larger ones, and that is where they preferred to nest. Grebes built nests earlier on the large islands, and these nests suffered less wind damage and predation and experienced higher hatching rates. There was no evidence that the clustering of nests had any effect on the grebes' success. The researchers concluded that exceptionally good habitat, rather than any particular advantage of nesting near others, caused Red-necked Grebes to become colonial. This sets them apart from most other colonial birds. I imagine that if good habitat consistently encouraged them into colonies, other colonial adaptations might follow.

Some colonies may not look like colonies at first glance. One of my favorite colonial nesters is the Purple Martin. Today, these lovely birds nest almost entirely in artificial homes, but they once nested in tree cavities or even in cacti. I myself have never had a garden big enough for a Purple Martin home, but many people put up colonial-looking homes with a few dozen chambers high on a metal pole or hang natural or artificial gourds in the hopes of attracting Purple Martins to nest in their gardens. It's a venerable tradition. According to references in a paper by Robert Allen and Margaret Morse Nice, Choctaws and Chickasaws hung gourds for martin nests,[18] and gourd-hanging was described as a practice in 1812 by Alexander Wilson (who wrote a nine-volume book, *American Ornithology*, before John James Audubon published, something that inspired Audubon's competitive streak throughout his life[19]). Apparently early Americans courted Purple Martins not for their insect-eating talents, but because they were thought to be highly aggressive toward vultures and hawks, which might otherwise feast on drying meat or skins or attack chickens. Interestingly, Charles Brown says this was a misconception and that it really was Eastern Kingbirds that attacked vultures and hawks. Another false belief was that Purple Martins eat mosquitoes—a rumor that he says was started by those who sold Purple Martin birdhouses.

Over time, according to Bridget Stutchbury, a former graduate student of Charles Brown's, Purple Martins largely abandoned naturally occurring crevices in favor of artificial homes across the eastern United States.[20] Stutchbury wondered how this transition might have affected the martins' colony sizes, and so she found a wild population that nested in saguaro cacti near Tucson, Arizona. She located cavities in the saguaros, mostly created by woodpeckers. Using a contraption involving fishing line, weights, rulers, and lighted mirrors, she counted the cavities big enough to fit a Purple Martin nest and observed the nests.

Stutchbury never found more than one breeding pair in a single saguaro, so no pair of Purple Martins had a neighbor closer than 328 feet (100 meters). Still, the nests were more clumped than random. Stutchbury wanted to know if birds spread across these comparatively distant nests still functioned like a colony, so she placed fake crows near nests and recorded whether the Purple Martins mobbed them. Indeed, they did. Purple Martins came from nearby nests and mobbed the crows, circling them, calling, and diving at them. More martins attacked when there were more nests in the area, indicating that the group functioned as a colony.

Since people put out more nest boxes than local martins need, we can see the density they prefer when there are plenty of options. Robert Allen censused Purple Martin colonies in artificial nest sites in Ann Arbor in 1940 and found twenty-two colonies spread over thirty-seven martin houses. Of the 761 possible nesting cavities, 191 were occupied, with 5 nests per martin house on average.[21] So the martins could have nested more densely, filling all the chambers in one martin house before a pair chose another house. Sadly, shortly after counting the martin colonies, in October 1943, Allen died serving with the US Armed Forces in New Guinea. Margaret Nice, one of the founders of the field of behavioral ecology, kindly took Allen's thesis and crafted it into a publication.

In natural cactus dwellings, as Stutchbury observed,

Purple Martins showed an advantage to nesting near one another: They fought predators together. This advantage holds for colonies in artificial nests, but there is another, more subtle advantage. Eugene Morton and his team worked on a single colony at Morton's home in Anne Arundel County, Maryland.[22] They found that older males returned to nest earlier in the spring than younger males did. I will call them mature males and second-year males. The mature males arrived at the site and set up their nests between April 2 and May 11. At this time, males fought for nest sites not only for themselves but, puzzlingly, to keep other nearby chambers empty of adult males. Why fight for a chamber you will not even use? Soon after this, females arrived, and the pairs began the business of mating and laying eggs. After the adult males' partners had begun laying eggs, the second-year males started to arrive and pair up with females that remained unmated. The mature males now allowed these pairings to use the empty nest chambers.

There was a good reason for allowing the second-year males to occupy nearby chambers: The older males mated with the female partners of the young males to such an extent that the youngsters sired only about 30 percent of the young in their own nests. By contrast, the older males sired 96 percent of the young in their nests. These older males were gaining an average of 3.6 eggs a season through mating with younger males' partners.

An alternative for the second-year males might be to nest alone, something Morton observed them trying three times. But females did not join them. Instead, they only chose males in colonies. Perhaps this was because of another social behavior that the Morton team observed: Second-year females snuck eggs into other females' nests. They found that 36 percent of all eggs in second-year nests had been laid by other females. The timing of this egg parasitism matched the breeding time of second-year birds and occurred well past that of the mature males and females.

Colony living may seem natural, for it is the way many of us live—well, perhaps with fewer shenanigans than we've observed among the Purple Martins! As is the case for many kinds of groups, colonies provide protection from predators. Colony members help one another by raising the alarm when predators arrive. They can also follow others that have spotted a good food supply. These advantages come with costs like reproductive competition, competition for resources, and increased parasites, costs that may explain why only about a fifth of birds nest colonially. The next chapter looks at the birds that take coloniality to an extreme: seabirds.

CHAPTER 5

Seabird Colonies

How to Rear Young by the Largest Larder on Earth

When seabird colonies come to mind, I think of birds flying in and out, soaring over rocky cliffs. I think of chicks, usually just one, looking very proper and upright at the entrance of its burrow or standing on a rocky ledge. I think of parents flying long hours far out to sea before they return to the one place on the guano-stained cliff they call home. I think of the boobies, albatrosses, cormorants, penguins, and frigate birds I saw on a trip to the Galápagos. I think of the island with bird-encrusted rocks that a Zodiac tour circled off the coast of Panama. I think of how, someday, I will travel to Norway or Iceland or up to Labrador to see even more seabirds and their huge colonies.

A key challenge governs nearly everything about seabirds: Though they feed at sea, they must return to land to breed. After all, no bird can incubate an egg floating on, flying above, or diving under the water. They must have a place to lay and warm their egg and then care for their chick. This means they have to make choices to optimize both a rich foraging ground and a safe nesting place.

Exactly how these birds balance their feeding and nesting needs varies across the more than three hundred seabird species. All seabirds fit into one of four orders: Sphenisciformes (penguins; 17 species), Procellariiformes (the tube-nosed seabirds: albatrosses, diving petrels, storm-petrels, fulmars, and shearwaters; 108 species), most Pelecaniformes (pelicans, frigate birds, gannets, boobies, and cormorants; 57 species), and most Charadriiformes (skuas, jaegers, gulls, terns, auks, guillemots, and puffins; 126 species).[1] Some seabird species forage hundreds of miles from shore, while others are more coastal. Some live and breed mostly in the tropical latitudes, others in the teeming waters of the

Arctic and Antarctic. Their coloniality, long lifespans, low reproduction rates, and distant fishing tell us how these 0.7 trillion birds (0.7 billion European) have mastered the waves around the globe, hauling in around 77 million tons (70 million metric tons) of fish, krill, and other invertebrates each year—nearly as much as humans harvest from the sea.[2]

I will begin in the cliffside burrow of a Manx Shearwater. These black-and-white birds soar far offshore over the Atlantic Ocean, landing only at night to tend their young on northern nesting grounds. The burrow I have in mind is one of about 320,000 on Skomer Island, Wales. Some of the colony's burrows are only a few feet deep, but others reach 20 feet (6 meters) deep and have side branches in which rabbits or storm-petrels rear their young.[3] Each nest will have a single occupant—like all tube-nosed seabirds, shearwaters lay just one egg a season. During the first week of a chick's life, one of its parents remains in the burrow to keep the chick warm, but otherwise it spends its time alone.

Why? Instead of remaining with their chick, both shearwater parents fly far out over the continental shelf hunting for sand eels, sprats, herring, and squid. After a successful foraging trip, they will float in rafts with thousands of other shearwaters not far from shore, waiting for nightfall before daring to visit their burrows and their precious chicks. This is how they evade diurnal predators. Once with their chicks, they will regurgitate fish caught from as far as 125 miles (200 kilometers) away right into their baby's mouth. Crops emptied, the adults are off again, leaving their single offspring in a dark hole until the following night.

In the next stage in their lives, young Manx Shearwaters, still fed entirely by their parents, will grow huge. By the time they are fifty days old and their feathers are almost completely formed, the young can weigh as much as one and a half times the weight of each parent. After a few more days of feeding, their job done, the parents will abandon their offspring and fly off to the coast of Argentina. For another week or so, the chick will stay in the burrow, sometimes venturing to the entrance or just outside to exercise at night. At seventy days old, the young are ready to leave their nest forever, off to join the ranks of their fellow marine creatures. In what seems like a crazy way to begin their independent lives, the young Manx Shearwaters walk to the edge of the cliff separating them from the water and tumble down. If it's windy, they might be able to take off flying. (Those that fall rather than fly easily survive.) Then they fly south, though they will not rejoin their parents. Young shearwaters must find their way and figure out how to fish all on their own. No wonder they need to carry some fat to tide them through their first days at sea.

Once they've left the nest, young shearwaters will not touch land for years. Relatively speaking, they are late bloomers who will not attempt to nest until they are five or six years old. When they finally do, they nearly always choose to return to their natal colony. There, their first task is a yearlong digging project, the hard work of creating their own burrow for future nesting.

This astonishing life story might seem extremely risky, with all that time spent flying over vast expanses of ocean in all kinds of weather, but Manx Shearwaters, outstripping most bird species, live for fifteen years on average—and sometimes many more. This longevity turns out to be a feature of many seabirds, and probably comes down to the lack of predators once they make it to the open ocean. Indeed, most of the Manx Shearwaters' mortal threats are land-based, often human. Humans once harvested hundreds of thousands of fat and tender young seabirds, calling them muttonbirds, and they've (not infrequently) introduced predators in the form of hungry rats and cats to the birds' breeding islands.

Ninety-eight percent of seabirds nest in colonies, according to David Lack, who is famous for figuring out the life cycles of birds.[4] The only exceptions are, according to Lack, a gull, a couple of murrelet species, and a handful of skua species. As we've already explored in the previous chapter, many factors can incentivize colonial life for birds: predator avoidance, proximity to food sources, food source identification (watching others to see where food lies), social functions like mate choice, and need for safe nesting sites.[5]

Let's return to the availability of nest sites, which seems especially relevant to marine birds: No matter how well adapted to life at sea, a bird must come back to shore to lay its eggs. To understand the role of nest sites, we can look for evidence that seabirds fight for prized nest positions within colonies. Central nests should fare better than edge nests. As we've seen, birds in the dense center of a group can suffer less predation. Seabirds tend to nest in places that are relatively free of mammalian predators, like small islands, promontories, or inaccessible cliffs, but there are still predatory seabirds, particularly gulls and skuas, as well as hawks, eagles, and falcons to worry about.

David Anderson and Peter Hodum looked at predation on Blue-footed Booby nestlings by Galápagos Hawks on the east end of Isla Española in the Galápagos Islands.[6] The hawks began to take booby nestlings as soon as the boobies were big enough to be left alone while their parents foraged at sea. Anderson and Hodum's colony lost more than half its thirty-nine nestlings in one season in 1984, mostly from the more isolated nests along the cliff. The researchers argued that the hawks choose these isolated nests because of the way they hunt, needing a few yards to land before they run up to the nest and grab a chick. Nests with more close neighbors come with the risk that adults might mob the hawks.

Back on Wales's Skomer Island, Tim Birkhead looked at the impact of colony density and nest location in Common Guillemots.[7] For decades, the guillemots had nested on a cliff face known as the Wick that was 200 feet (60 meters) high. Because the nests were inaccessible to humans, Birkhead had to rely on photographs of the cliff face for much of his study. One thing the photos showed was that the birds preferred nesting on the narrowest ledges, only about a foot wide. Birkhead compared what he saw with photos taken of the same colony in 1934 to understand how the guillemots' nesting patterns had changed over time. In the forty-one years preceding his study, the colony had shrunk considerably, and nests were no longer packed as densely. This sparser colony had reduced breeding success, most likely because the nests were more exposed to gull predators.

Atlantic Puffins are a colonial seabird I would love to see someday. David Nettleship studied Atlantic Puffins on Great Island, in Witless Bay, on the southeast coast of Newfoundland.[8] Not quite a mile long and half a mile wide at its broadest, Great Island is edged by granite cliffs and deep bays. It looks a bit like a decaying molar shoved up out of the ocean. Yet, with barely a cove to land a boat, this is home to more than one hundred thousand pairs of Atlantic

Puffins, who happily excavate their burrows in the steep meadows atop the cliffs. They share the island peacefully with Leach's Storm-petrels, Black-legged Kittiwakes, Razorbills, Common Murres, and Black Guillemots. Things are less peaceful when it comes to the Herring Gulls and Great Black-backed Gulls that prey on puffin chicks.

Nettleship divided the nesting site into a grid of 20-foot (6-meter) squares and, watching with an assistant at various times of day from a sheltered blind, kept track of when and where the puffins built. The scientists color-banded the puffins, checked what kind of food they brought back to their chicks, and measured nesting timing and success. I was most interested in their data comparing the nests on the slope approaching the brink of the cliff and those on the more level ground behind it: Most burrows were on the slope; the less desirable level areas were chosen for nests only after the slope was fully occupied. Many more chicks fledged successfully on the slope than on the level ground, where gulls were more likely to take chicks and to accost parents returning with food in their beaks. Adult puffins that launched from the slope were able to take flight within a second, but those trying to launch from level ground had to run first. It took them much longer and made them more vulnerable to gulls. This made the slope-side burrows better spots for fledglings. Nettleship went on to compare egg and chick survival on Great Island to that on the smaller Funk and Small Islands, which had few gulls (Funk) or no gulls (Small). He found that puffin breeding success was much lower on Great Island: 37 percent overall on Great Island compared to 87 percent on Funk Island and 93 percent on Small Island. The chicks were also much heavier at the time of fledging on the gull-free islands.

In another landmark study of nest position and success, John Coulson looked at the Black-legged Kittiwakes in Northumberland, UK.[9] In this gull species, males choose nest sites on cliff ledges. Coulson found that his birds competed hard for central nests, in which birds had larger clutches and higher hatch success, fledged more young per pair, and suffered 60 percent less nestling mortality than birds nesting on the colony edge. When Coulson made new nesting sites available in the middle of the colony, they were quickly occupied, while similar sites added to the edge of the colony stayed empty. Interestingly, Coulson found no evidence that predation caused these differences. So why did the birds compete for the central nest sites? Perhaps the birds in the center are stronger and able to win prized central locations, which are safer from predators even though there are no longer such predators.

Seabird colonies range dramatically in size. Black Guillemot colonies number no more than 150 birds, while Manx Shearwater and Atlantic Puffin colonies can top 100,000. Leach's Storm-petrel and Common Murre colonies can have a *million* or more breeders. This variation likely results from differences in feeding ranges.[10] Birds that can travel farther out to sea to forage can congregate over larger areas. In one classic study, Philip Ashmole went on an expedition in 1961 with the British Ornithological Union to Ascension Island, south of the equator in the mid-Atlantic.[11] There he focused his work on several species of terns, including the Sooty Tern, nicknamed the Wideawake Tern for its incessant calls, and the Black Noddy, whose head bobs.[12] Ashmole said that because there was space in the area that was not used for nests, nesting sites must not be the limiting factor for bird numbers—rather, it must be food. He thought that the waters around a colony could become depleted, a phenomenon subsequently referred to as Ashmole's halo.

Robert Furness and Tim Birkhead looked for Ashmole's halo in Britain and Ireland, where some had thought that seasonality would make seabirds' food superabundant during the breeding season and therefore unlikely to limit colony sizes.[13] But it turned out that Ashmole's halo still applied for Northern Gannets, Atlantic Puffins, European Shags (cormorants), and Black-legged Kittiwakes. I guess even in the northern Atlantic food can be limiting.

More evidence for a food-depletion halo around a colony involved Double-crested Cormorant colonies in Maritime Canada.[14] Prey fish numbers there were lower in bays near active colonies than in more distant but otherwise similar habitats. And larger colonies of Northern Gannets grow more slowly and are most distant to the best food patches, requiring birds to fly farther on each foraging trip.[15] It appears there is evidence for fish depletion near seabird colonies in both temperate and tropical waters. This means that seabirds should nest where there are the most fish nearby.

From the shore or even from a boat far out at sea, the ocean looks uniform, an endless expanse leavened only by subtle changes from gray to blue, but with depths of purple or brown and a froth of white. It continues featureless for as far as one can see, at times flatter than anything on land. Even the greatest storms bring only passing contour changes. If the seas were as uniform as they appear, then a seabird's choice of a nesting site could be based entirely on terrestrial features. The bird could choose the steepest cliffs or the remotest predator-free islands, for example.

But the uniform surface of the ocean belies the variability that lies below. The fish, squid, krill, and other invertebrates that seabirds eat are far from evenly distributed. There is a vertical gradient from near the ocean's surface, where light and oxygen penetrate and life flourishes, down to the depths, where decaying bodies and their nutrients settle. The greatest density of organisms occurs when the rich soup that has settled deep in the ocean is brought back to the surface and living organisms can recapture its nutrients.

Ocean currents mix the waters of the world. The processes that generate them are complicated. The Coriolis effect, caused by Earth's rotation, results in gyres as vast as entire ocean basins moving along the coasts in a clockwise direction in the northern hemisphere and a counterclockwise direction in the southern hemisphere. Other factors include water temperature, trade winds, ocean-floor topography, and salinity. For our purposes, what is important is that only some regions have currents that bring the remains of past life up from the depths.

The current I know best is the Gulf Stream, which flows north along the eastern coast of the United States, across the northern Atlantic to the European coasts, then splits north into the Norwegian Current and south into the Canary Current. The Gulf Stream is what keeps Europe so much warmer than similar latitudes in North America. Another well-known current is the Humboldt Current, which flows north along the west coast of South America, keeping Chile, Peru, and Ecuador much cooler than they would otherwise be. The Humboldt Current causes coastal upwelling of deeper water, which feeds surface organisms year-round in Peru and in spring and summer in Chile; it makes the waters so cool that penguins live in the Galápagos off Ecuador in equatorial waters.

The Humboldt Current is one of the world's great fisheries, and it is heavily fished for sardines, anchovies, jack mackerel, tuna, swordfish, hake, and squid. The upwelling that makes this current so productive fails in El Niño years, when the surface waters of the central and eastern tropical Pacific Ocean warm, stilling the usual currents and preventing nutrients from rising to the surface. El Niño is connected to an atmospheric condition called the Southern Oscillation; together they are called ENSO. In an ENSO year, the trade winds diminish and rain patterns change over much of the planet. The resulting lack of oceanic turnover is disastrous for the birds and people who depend on the fishery. As I write this, we are headed into an El Niño period, one that shows every sign of being historically strong.

Now that we have an idea of the variability underlying the ocean's surface, we can consider whether food availability predicts colony size within a single geographic region. A thorough study of seabird nesting locations was done along the coast of Norway, home to four million breeding seabirds. Hanno Sandvik and his team examined the distribution of twenty-seven seabird colonies on the Arctic coast from 67 to 71 degrees north latitude along the Barents Sea.[16] Of the colonies they covered, the largest was at Røst, with more than a million birds, mostly Atlantic Puffins, but also Common Guillemots (Murres) and Black-legged Kittiwakes. Next were Gjesværstappan and Vedøy colonies, each with nearly half a million birds, again mostly Atlantic Puffins but also Razorbills, Black-legged Kittiwakes, and Common Guillemots.

Sandvik and his team sought to determine whether fish location in the waters off Norway explained the location of the colonies. After all, nesting habitat was unlikely to be limiting, since that entire 750-mile (1,200-kilometer) stretch of coast was potentially good for nesting, with cliffs, islands, peninsulas, and turf behind the cliffs.

Sandvik's team did not look at the seabirds' common prey fish—capelin, herring, and sand eels. Instead, they looked one trophic level down, at the plankton that serve as food for the fish. By definition, plankton are free-floating organisms that are too small to move far on their own. Passive forces based on ocean currents and characteristics of the seafloor will determine where these organisms are. Where the continental shelf is narrow, the current flows closer to land, and where the shore is complex, with small islands and peninsulas, nearly stationary eddies further concentrate small crustaceans, fish eggs, and larvae. Presumably the fish, then the seabirds, will follow.

Indeed, Sandvik and his team found that seabird colonies were concentrated at places where plankton (and thus fish) were most abundant. The investigators took into account the likely distance seabirds would forage. It turns out that colonies tended to be between 3 and 5 miles (5 to 8 kilometers) from the best foraging locations. But the birds often fly much farther—even as far as 15 miles (24 kilometers). Thus, colonies are closer to the best foraging locations than would be predicted by chance. Interestingly, Sandvik and his team noticed that some locations with high numbers of plankton did not have any nearby seabird colonies. They attributed this to higher variation in plankton at those locations. After all, if birds nest in a colony every year, they will want to avoid places where the food sometimes disappears. The birds not only need to be close to plenty of fish; they need that fish supply to be dependable.

I cannot conclude a chapter about seabirds without mentioning the most adored seabirds of all. The Sphenisciformes, or the penguins, are an order of birds that dominate the seas around Antarctica, though they evolved in New Zealand. Collectively, penguins make up 90 percent of the avian biomass in the Southern Ocean, from 60 degrees south latitude to the shores of Antarctica.[17] Most species' breeding takes place during the summer—only the Emperor Penguin breeds in winter, huddled on sea ice, incubating eggs and chicks on fathers' feet. Penguins rear their young in huge colonies, but they are monogamous and mate with the same partner from year to year (except for Emperor and King Penguins, who have a higher degree of partner turnover, probably because they do not have a fixed nest site to return to).

David Ainley and his team looked into colony distribution for the three species of penguins in the Pygoscelidae family, the Adélie, Chinstrap, and Gentoo Penguins, in the western Antarctic Peninsula.[18] Ainley's study was not small;

it was based on nearly 600,000 pairs of Chinstrap Penguins, about 125,000 Adélie Penguins, and about 24,000 Gentoo Penguins. They tested the hypothesis that prey depletion around colonies would limit colony size, as per Ashmole's halo. First, they had to determine the birds' foraging range. Foraging Adélie Penguins averaged 78 miles (126 kilometers) away from the colony per foraging trip, while Chinstrap Penguins averaged 56 miles (90 kilometers), and Gentoo Penguins averaged 37 miles (60 kilometers). No wonder the Gentoo Penguins I saw swimming underwater at the St. Louis Zoo moved so fast with the slightest twitch of a flipper. Members of their species were used to going much, much farther than their zoo habitat permitted.

The researchers did not find evidence that prey depletion was limiting colony size in these rich southern oceans where prey is superabundant. Instead, colonies were clustered, with smaller colonies near larger ones—a sort of suburban pattern that's likely the result of penguins preferring to nest near their natal colony but not always finding space there.

Seabirds have few young, and they invest mightily in each hatchling. As noted earlier, the petrels, albatrosses, and shearwaters in the Procellariiformes never lay more than one egg a season, and birds in the other seabird orders lay at most two. They do not begin to breed until they are several years old, and breed only once a year at most. They incubate their egg for one to three months, depending on the species, and then provide extended care for the chick (though many species do not care for young at all once they have left the nest). By contrast, songbirds like thrushes or finches breed in their first year, lay several eggs at a time, produce a new clutch if the first one is lost to predators or the weather, and minimize the time both eggs and young remain in the nest.

This seabird way of life has been effective for thousands of years, but today seabirds are declining because of human activity.[19] Their nesting habitats, especially in the islands of the tropics, where they overlap more with humans, are often destroyed,[20] and humans diminish their food supplies by hauling ever increasing shares of fish and cephalopods from the oceans, using nets and lines that further risk entangling and drowning birds.[21] Overfishing and pollution are major problems, particularly for penguins, petrels, shearwaters, cormorants, and pelicans,[22] while climate change destroys the habitat necessary for seabird survival by changing up-welling currents and melting polar ice floes.[23]

Human-induced seabird troubles are not all recent. David Steadman documented prehistoric reductions in native birds on the topical islands of the Pacific Ocean by comparing the birds there now with those whose bones were found in archaeological sites. For example, the fossil record of seabirds on Easter Island indicates that there used to be twenty-five species. Of those species, one is now extinct and twelve to fifteen are no longer in the area. Only the Red-tailed Tropicbird still breeds there. The extinction happened from 1,500 to 550 years ago, just when humans first colonized the island.

In the Marquesas Island of Ua Huka, a two-thousand-year-old archaeological site called Hane yielded the bones of shearwaters and petrels from about seven species, none of which still nest on the island.[24] The archaeological remains indicate that the vanished birds were once a major part of local humans' diet. This pattern of human arrival coinciding with the extirpation of endemic birds repeats itself all through the South Pacific islands. Steadman put land and seabird data together and estimated that "an average of 10 species or populations have been lost on each of Oceania's approximately 800 major islands."[25]

In addition to the direct harms from humans, the animals that humans have introduced can be catastrophic to seabirds. Holly Jones and her team studied the impact of one of the most ubiquitous of human commensals: rats.[26] All over the world, in the colonies of seventy-five documented seabird species, they found rats, which preyed on seabird eggs, chicks, and adults. Jones and her team found that rats had successfully invaded islands from the tropics to subarctic tundra, causing devastating effects on bird populations (killing off the seabirds entirely in ten cases). The storm-petrels, auks, and murres suffered the most.

Ultimately, climate effects may present the biggest threat to seabird populations. Events that cause massive die-offs in seabirds are called wrecks.[27] One such event happened between the summer of 2015 and the following spring, when around sixty-two thousand Common Murres washed ashore on Pacific beaches, from Alaska to California. Since most dead birds do not wash ashore, John Piatt and his team estimated that the actual death toll was one million, concentrated in the Gulf of Alaska, where Common Murres are the dominant fish-eating bird. The dead birds that were recovered were emaciated, and researchers think starvation killed them.

How? Piatt and other researchers attributed this wreck to the strongest marine heat wave on record. Amid water temperatures never before seen, there were cascading effects on marine life, from tiny phytoplankton up through the fish upon which marine birds depend. In fact, it wasn't only the

Common Murres that were affected. There were also die-offs of Cassin's Auklets (which eat plankton), sea mammals in Southern California, and baleen whales in the Gulf of Alaska.

Climate change is also having marked impacts on penguins.[28] Species that depend on ice, like the Emperor and Adélie Penguins, have reduced their ranges and begun adjusting to climate change by altering what they eat.[29] As ice shelves contract, ice krill diminishes, and the penguins switch to eating silver fish, for example. Penguins are vulnerable to a warming and more variable climate in the southern oceans. They may be an extreme example of the tragedy of climate change, but all seabirds are impacted.

Seabirds ring our oceans with life, soaring through the air, flapping at the water's surface, or gliding effortlessly through the water. They compete with us for the world's fisheries, and their presence tells us where sea life is most abundant. Just like us, they live mainly in colonies of astonishing size. Understanding these colonies and the remarkable lives of seabirds is challenging, but it is well worth the effort as we strive to conserve and protect them.

CHAPTER 6

Leks

Where Males Dance and Females Choose

It was still misty when we saw an Attwater Prairie-chicken and heard its melodious booming and chattering. Drumming its feet, as if dancing in place, the bird inflated its yellow air sacs—they might have been mistaken for cheeks, were they not so large. Then, nearby, we spotted another, and another. These were males, each dancing on his own tiny court, awaiting a female's decision. Would she choose him? For their part, the females were experts at nonchalance, walking calmly among the males before selecting one for a quick mating. To us, the males all looked alike, but to the females they did not. Many of the females chose the same male, often the one whose court stood on the hard-won scrap of ground near the center of the lek. A lek is a place where males congregate to attract and mate with females, usually by displaying on a tiny territory. That is the extent of their interaction with females, as lekking males provide absolutely no parental care.

On this morning, cattle were on the lekking ground, their skinny legs forming a backdrop to the dancing males. We were told that the cattle were necessary to maintain the prairie grasses, but I wonder if the cattle contributed to the critically endangered status of Attwater Prairie-chickens. Only about two hundred are left in the wild.

Attwater Prairie-chickens are one of a small number of bird species that form leks. Emily DuVal, my friend and former student, and current president of the Animal Behavior Society, tells me that the word *lek* comes from the Swedish word for "play." Apparently, some early observers of lekking males in Sweden thought the birds' pronounced displays and dances bordered on the ridiculous and concluded the behaviors must be play.

It is a sort of game, though a high-stakes one. In leks, males compete vigorously for the best spots so that the females, who visit the arenas only briefly to shop for a suitable mate, are more likely to choose them. Once they have mated, the females depart to build their nests elsewhere. Fewer than a hundred bird species form leks—mostly shorebirds, grouse, hummingbirds, cotingas, birds-of-paradise, and manakins—but leks are so puzzling and flamboyant that they have received a lot of attention.[1]

Leks are thought to make it easier for females to judge males and mate with the best. They are most common when individual females forage across huge areas, so males cannot easily find or monopolize either the females or their food sources. Males group in leks in locations where nesting or foraging females are likely to be encountered.

Some scholars consider leks a biological paradox, because females choose as if it matters which male they mate with, even though all they get from him is sperm.[2] If they were choosing a male that could provide something material to them, female choosiness would be easier to understand. We can think of it as a more extreme version of females choosing to mate outside the pair bond (as we will see in the next chapter). But here there is no infidelity, because there is no pair bond to betray. So in birds that form leks, it's all about females carefully picking the father of their few precious eggs (the males can produce enough sperm to mate with many females).

Others think there is no paradox—at least, that leks make a certain sort of sense. As we all do, male birds vary genetically, and dancing for long periods in an arena in which they have to fight for territory helps sort the champions from the weaklings. All the females need to do is go to that arena and mate with the top male, even if each must wait her turn.

So, what *is* the lek paradox? It has to do with the slow pace of evolution. For if the top male is always the one to mate, his genes would spread through his many descendants. Ultimately, there would be no variation in the traits that mattered, and so females would be unlikely to find that one male was genetically better than another. Gerry Borgia first called this the lek paradox and applied it to the bower birds he studied in Australia.[3]

But since females act like there is a difference among males, there probably is. Maybe some force maintains variation among the males. One such force might be disease, since disease might target the commonest genotypes. Parasites and diseases maintain variation because they constantly evolve to better attack individuals with the most common genotypes. That means the identity of the fittest males could continuously change over the generations, as formerly rare genotypes become more common and thus more susceptible to disease. This idea is called the Hamilton-Zuk hypothesis in honor of its famous authors.[4]

The paradox of female choice is a big question, and we need not solve it here. What is important for our story is that females clearly *choose*. Whatever the reason, they compete to mate with the male with a preferred position in the lek and the best display, both of which are costly for him to maintain.

One of the best-known lekking birds is the Greater Sage-grouse. They were first described by Lewis and Clark on the explorers' famous 1805 westward expedition, which left from my hometown, St. Louis, Missouri. Back then, the Greater Sage-grouse was common, but now these birds are classified as "near threatened" because their habitat has been overgrazed, cultivated, or buried under subdivisions.[5] More recently, habitat has been lost to ever-larger fires burning east of the Rocky Mountains, as well as to oil and gas development. Wyoming has 40 percent of all surviving Greater Sage-grouse.

The Greater Sage-grouse is a bird of the sagebrush ecosystem, associating with several kinds of sagebrush and the other plants found among dry, cold fields with gullies running along the waterways. They form their leks in the open, though close to more densely vegetated areas suitable for nesting. At dawn and dusk on the lekking grounds, male Greater Sage-grouse show off for the females. The males strut about, tails spread, filling their esophageal pouches with air until they look like nothing so much as large mammalian breasts, inflating and deflating. The deep plopping sounds their syrinxes make as the air is expelled can be heard half a mile (0.8 kilometers) away. I recommend finding videos of them on the Cornell Lab of Ornithology site to see their remarkable behavior.

John Scott undertook one of the earliest studies of Greater Sage-grouse mating behavior on the plains near Laramie, Wyoming.[6] There he observed leks, which he called "strutting grounds." Scott went into a shelter that hid him from the birds before they arrived and stayed until they were done strutting, around eight o'clock in the morning. Females arrived at the lek a few weeks after the males and spent a lot of time checking the males out before actually mating during the second half of April. Scott watched them from mid-March to mid-June, observing them for at least fifteen days.

Scott concentrated on a large lek that was 200 yards (183 meters) wide and half a mile (0.8 kilometers) long. At the height of the season, as many as four hundred males occupied the area. Females were harder to count because they came and went as they shopped for a mate. The same site was used year after year. One lekking ground Scott observed continued to be used even after a road had been built right through it, something that unfortunately led to a lot of road-killed males, as researcher Gail Patricelli (whom we will meet soon) told me.

Though the lek was quite large, mating occurred only in four small areas, each just a few yards wide. Always attending these spots were a group of hens and two males, one that Scott called the "master cock," who marched and displayed vigorously on his court, and a less flamboyant "sub-cock." Other males hung about at a safe distance.

The bird Scott called the master cock monopolized the mating opportunities at each spot. Scott saw one male mate with twenty-one hens in a single early morning. Two mornings later, a male in the same arena, likely the same male (Scott could not say for sure, because the birds were not banded and he had not yet learned their unique feather patterns), mated with fifteen hens. Another mated with five hens in just nine minutes. Even the most vigorous birds could not maintain such a pace forever, though, and slowed down later in the morning. Scott saw one male mount his fourteenth female of the day but finally proved unable to consummate the act (Scott could distinguish successful matings because, afterward, females shook themselves vigorously and carefully preened their feathers, their version of jumping into the shower before carrying on with their day). When mating attempts did not end with satisfied preening, the resident sub-cock, the second in command, would dart in and take the master cock's place, though he was often attacked by the master cock for doing so.

Scott attributed the birds' predawn and early-morning mating behavior to a fear of Golden Eagles, which were, during his observations, the main predator of the Greater Sage-grouse. He said the eagles seldom flew over the lekking grounds before sunrise but overlapped enough with Greater Sage-grouse activity that many were lost to eagles.

About thirty years after Scott's observations, R. Haven Wiley studied Greater Sage-grouse in the same location.[7] Sadly, the grouse had declined in the interim. At one lek, Wiley found only 154 males in attendance, down from the 355 observed by Scott. The number of mating centers—the tiny territories where most of the mating happened—had declined, too.

Early every morning, Wiley observed the grouse from a hidden blind. He did not band the birds but learned to tell the males apart from their tail shape and the pattern of their undertail coverts, which show when the tail is fanned. (Patricelli calls these "butt prints.") Wiley did not see quite the level of mating monopolization that Scott had. While Scott had found 75 percent of the matings went to the dominant male, Wiley found that the dominant male got only 53 percent of the action.

Mating is the whole point of the lek, so I wondered exactly what the females do when they arrive. Females first arrive at the leks in early spring in groups of five or fewer birds, but by the time mating season begins in mid-April, there could be a hundred or more females visiting a mating center simultaneously. Once in the area, they shop around, often visiting several leks and mating territories within larger leks. Once they make their choice and mate, they leave the lek and are not seen there again. This one mating is enough to fertilize all the eggs they will lay, usually six to eight in a season.

Given that females look at multiple leks, they seem to be choosing the one male that they judge to be best—presumably remembering all the males they visited at other leks and comparing them in their minds. But what makes a male the best? Is it being strong enough to hold the dominant spot at a mating center? Perhaps it's related to the loud plopping sounds that can draw females from far away? It is important to remember that mating is most common in the dim light of earliest morning, when visual displays will be subtle.

Rob Gibson and Jack Bradbury tackled this topic in a study that considered males' anatomical characteristics, display performance, position in the lek, and fighting status.[8] At a small lek in California, near Crowley Lake in Mono County, Gibson and Bradbury put bands on twenty-seven males so that they could definitively identify each bird. Then they watched from a blind on a small rise, which gave them a great view of the area with spotting scopes. Gibson and Bradbury observed much less monopolization of matings than earlier investigators had found; of their twenty-seven males, thirteen mated—six mated once, four mated twice, and three mated three times. Maybe this small lek was atypical, but it still yielded information on what distinguished the males that mated from the fourteen that didn't. Gibson and Bradbury found that the key differences were simply time spent on the lek and features of the display. Dominance and territoriality did not seem to be as important. Clearly Greater Sage-grouse are complicated, with females all mating with a few males in some places while choosing different males in other places.

Some researchers are skilled at technical wizardry and use this to extend our knowledge of animal behavior. One of them is Gail Patricelli. I know her best from the Winter Animal Behavior Conference, held every January in Steamboat Springs, Colorado. We are mainly there for the talks and scientific conversation, and some of it takes place on the ski slopes, where Gail is the most proficient skier, leaving the rest of us in the snow as she weaves expertly down the hardest runs.

Patricelli and Alan Krakauer decided to use a robot hen to better understand how males respond to females. The robot was a stuffed female Greater Sage-grouse that moved along G-scale model train tracks.[9] The head could move from side to side, and the body could rotate completely to face the courting male. The robot contained a video camera, a microphone, and an audio recorder. Patricelli and Krakauer laid the tracks out from the blind straight to the territory of the male of interest, about 25 yards (23 meters) away, then camouflaged them as best as possible.

The advantage of a robot is that the researchers could control exactly what it did and then see how males responded. Patricelli and Krakauer tested twenty-eight males, choosing times when no other females were there to avoid confusion. They were particularly interested in whether the males faced a trade-off between display quality and duration. If they displayed longer, would their display quality decline? Because the robot female did not get bored or wander away, they could test the males for as long as they wanted. They looked for struts, a swish of feathers against the breast, and paired plops, also called pops, separated by a whistle.

Patricelli and Krakauer had the robot stop three times at different distances from each male they tested. They then recorded the male's response via strut rate and the interval between pops. They found that the males that had higher strut rates toward the robot also had the highest mating success with the non-robotic hens. This indicates that females like a fast strut rate. There was no relationship between pop interval and mating success. Males with high reproductive success seemed to be able to keep up vigorous strutting as long as the female was close.

Most studies of Greater Sage-grouse mating are at leks. But males that do not win at leks might try to approach females away from leks. Katy Semple and her team decided to see if this was the case by following females with radio transmitters and collecting eggs to see who fathered them.[10]

The Semple team found that the genetic data from the eggs generally matched what they had observed, but some young were fathered by males not present at any lek. There is a little more to Greater Sage-grouse mating than is apparent from lek observations. Yet it is still true that a male that displays at the place in the lek the females favor will have more young than others.

Another lekking bird is even more complicated than the Greater Sage-grouse. The Ruff can be found breeding from northern Europe to the whole of northern Asia. Ruffs winter in sub-Saharan Africa, as well as in other southern coastal areas and the Malay Archipelago, and since they must fly between breeding and wintering grounds, they might be seen in wet habitats nearly anywhere except the Americas.

I once saw captive Ruffs in breeding cages maintained by David Lank, who goes by Dov, at Simon Fraser University, before he moved them to a bird institute in Germany in 2021. The Ruffs had long, pink-orange legs, to which Lank had attached bands of yellow, green, orange, blue, or white so that researchers could identify them. Their long beaks were dark toward the tip and pink toward the base. Females had feathers edged in white, but were otherwise dark or light brown, with a lighter chevron in the dark area. Some males had spectacular black, rusty, or white neck ruffs and head tufts of loose feathers they spread to display.

Ruff leks are quiet places, devoid of song. Lank considers plumage variation of the Ruffs to be their silent song.[11] The males' head tufts vary in color intensity, and their ruffs can vary in pattern, while the skin of the wattle runs from deep red to yellow. When Lank manipulated the birds' colors with markers, he found that the Ruffs were treated as strangers and with aggression, supporting the idea that their patterns served as identifiers in the same way that songs can identify birds.

Like the Greater Sage-grouse males, Ruff males form leks that consist of small arenas for display in the breeding season. Alida Hogan-Warburg studied Ruff leks in damp Dutch meadows, where males arrived from their wintering grounds in mid-March, followed by the females in about the second week of April. The males and females did not stay all the time on the lek but instead alternated between the lek and foraging in the meadow.[12]

Ruff display arenas are only about a foot (30 centimeters) across. Hogan-Warburg recognized that there are two main kinds of males within them. Independent males—the ones with black or rusty-colored ruffs and hairdos—defended these small courts, leaving only for short foraging trips that became rarer and rarer during the height of the mating season. They would compete for the best court near the center of the lek, pushing younger and less experienced males to less desirable spots near the lek margins.

Satellite males are quite different. They are slightly smaller, on average, and generally have white or rusty neck ruffs and head tufts (never black). They are much less aggressive and do not fight to hold territories of their own. Instead, they are recruited by the territorial males as temporary display partners. Satellites are rarely attacked by independent males, so maybe it does not make sense to call court holders "independent"—they recruit satellite males to help them woo females. Don't worry, there can be something in it for the satellite males, too, as I will explain.

Females in Hogan-Warburg's study were more likely to mate with independent males if a satellite was present on small leks, but less likely to do so on large leks. For example, Hogan-Warburg found that at the large lek at Polder Oosterwolde (about an hour's drive east of Amsterdam), there were fifty-three copulations over the course of nine days. Over half of the resulting chicks were fathered by just three of the nineteen independent males. Eight independent males never copulated. Meanwhile, only two of the twenty-three satellite males copulated, generally when the independent male was chasing someone else away. Small-lek copulations were more evenly divided. At the small lek at Schiermonnikoog (an island farther north), there were 128 copulations over the course of forty-eight days. Nearly three-quarters were fathered by the five independent males, the rest by three of the five satellite males. Satellites seemed to be following a low-effort, low-reward behavioral plan, but it was not optional.

Hogan-Warburg speculated that the difference between independent and satellite males came down to a genetic polymorphism, an idea that was eventually confirmed by Dov Lank and his team.[13] To get pedigree data, Lank reared Ruffs in captivity, housing females with single males. His team found nearly all sons of independent males grew into independent males, and half of the sons of satellite males grew into satellites. With this and additional data, they determined that there was a genetic basis for being an independent or a satellite. There was likely to be one gene that could take one of two forms. If there was an allele for satellites, the male would be a satellite. They speculated that whatever gene was causing the difference, it did not change what females looked like.

In the mid-1990s it was not possible to determine what the genetic basis for this sorting might be. The genomic revolution changed that. Two big groups of researchers discovered the mechanism and published their findings simultaneously.[14] It turned out there was an inversion on chromosome 11 in the satellite males. An inversion happens when a piece of a chromosome breaks off and is accidentally repaired backward, so the starting place becomes the ending place along that piece of the chromosome. In the Ruffs, the length of DNA that this happened to had just over one hundred genes on it. Once this piece was flipped, the genes on it remained functional except for those at the actual break site, but none of the genes in the inversion could evolve as easily or recombine because they no longer matched up with their partner chromosome. This explains why independent and satellite males could look so different and have such different strategies. The whole inversion acted like one single trait but contained enough information to impact plumage, size, and behavior. An individual that had both of their two chromosome 11 copies flipped could not develop, making this inversion lethal when homozygous, in technical parlance—meaning that birds with the flipped chromosome also had one normal chromosome. As is often the case with inversions, the few genes disrupted by the flip cause the behavioral changes, and they are often related to hormones and brain functions.

Joop Jukema and Theunis Piersma noticed that, just as there were two types of males with different body sizes, there seemed to be two different sizes of female Ruffs. They decided to explore this. Working in Friesland in the Netherlands, they banded more than a thousand birds in the spring of 2004.[15] They weighed and measured the birds, collecting a tiny drop of blood so that they could do molecular tests.

Joop Jukema is not your typical ornithologist. He is a Dutch potato farmer who has a passion for birds, first Golden Plovers and then Ruffs. He says on his website that science is a wonderful thing if you do not have to make your living at it. He seems to have spent all his free time in the service of birds, winning quite a few honors in the process.

Jukema and Piersma discovered that some of the birds that had the form and color of females turned out to be genetic males. In fact, we now know there are *three* types of Ruff males. This third form, which Jukema and Piersma called "faeders," look just like females, with rich brown feathers edged in white, but secretly boast testes as large as the other forms.

Faeders are nothing like the flamboyant independent males with their big ruffs and darker feathers. Nor are they like the white satellite males. They are a different form entirely. They are uncommon, making up no more than about

1 percent of all birds. In Jukema and Piersma's observations, faeders' female disguises seemed convincing enough to fool independent males into attempting to copulate with them (though this could have been part of the normal male behavior of mounting each other, since faeders were often on top).

But wait! I have talked about one chromosomal inversion and yet there are *three* different morphs. It turns out Ruffs are even more complicated. The chromosomal inversion I've already described happened about four million years ago and resulted in the faeder males. But then, about seventy thousand years ago, two additional inversions turned two small areas of the inverted region *back* to the original orientation found in the independent males. This yielded the satellite males, which are closer to independents in that they have ruffs, they display (though they do not fight), and they grow white or at least pale display feathers, never black. Many have white ruffs with rusty head tufts.[16]

How the faeders interacted with the independent males for the millions of years before the satellite inversion occurred is lost to the mists of evolutionary time. But once the inversion happened—just an evolutionary blink of an eye ago—it seems that cooperation between the satellites and the independents evolved quickly. It is an unbalanced kind of cooperation, wherein the independents get most of the matings while the satellites help attract females and get a lower proportion of the matings.[17] But the faeders and satellites get just enough matings that the trait does not go extinct, maybe because their carriers do not have to fight for territories, and maybe because the trait is also in females who do not pay an apparent cost.

Another thing worth understanding about Ruffs is that, unlike the Greater Sage-grouse females, Ruff females often mate with more than one male.[18] Since females in a lek mating system receive only genes from males (not any help or access to territory), the standard view is that the female picks the best male, mates with him, and then leaves. If she mates with more than one male at a time, she is effectively letting the sperm battle it out in her reproductive tract, selecting the best that way. The importance of this sort of sperm competition to Ruffs is reflected in the fact that Ruff sperm are longer than those of any other shorebird—long sperm cells have a better chance at fertilizing a female's eggs.

To see if females mated with more than one male, Dov Lank and his wife, Connie Smith, along with other collaborators, worked in Finland, southeast of Oulu, making daily observations from blinds through the breeding seasons in 1987, 1989, and 1990. They collected blood from males and females on nests near the leks, using it to look at the parentage of the thirty-four broods that they sampled, either young or unhatched eggs. Half of the broods had two or three different fathers. This high number of multiple matings had to have been from multiple visits to a lek, since it did not match what they saw when a female visited a lek once.

To really understand the scope and drivers of multiple paternity, one would need to sample a lot more nests, which is a hard thing to do in most lekking species because females are drab and secretive and may nest far from leks. So, we turn now to Lance-tailed Manakins in the Pacific lowlands of southwestern Costa Rica. Lance-tailed Manakins are tiny compared to the huge Greater Sage-grouse and substantial Ruffs. Emily DuVal says I could fit one in my pocket, though of course I never would.

Male Lance-tailed Manakins are beautiful birds. They have a black body, a baby blue cape, and a brilliant crimson crest that they flare into a spiky crown when excited. Their two central tail feathers stick out in a narrow point—the lance that gives them their names. Females are duller, a camouflaged olive green with a pale belly.

I have known DuVal since 1993. I have a special fondness for her because she was an undergraduate at Rice University when I was teaching there. She studied the Great-tailed Grackles right outside our building with my colleague and friend Kris Johnson. DuVal was even in Wiess College, a residential and activities college, with us and in the small group of freshmen my husband and I hosted right before classes started. I was particularly proud when she became president of the Animal Behavior Society in 2024, taking over from the previous president, Vanessa Ezenwa, who was in that same freshman group at Rice University, studied wasps in my lab, and now studies disease ecology in Cape Buffalo and other South African ungulates.

DuVal studies Lance-tailed Manakins on Boca Brava, an island not far from the city of David on the Pacific side of Panama near the Costa Rican border. The DuVal study site is on a peninsula of the island closest to the mainland. It is about 100 acres (40 hectares) of secondary growth with some homes and sheds. It is bordered by cow pasture, which keeps this population fairly distinct, since the Lance-tailed Manakins tend not to cross the pasture. DuVal identified around forty display arenas in the study area, spaced about 111 yards (101 meters) apart. This proximity means birds at one arena can hear and sometimes see their neighbors.

Lance-tailed Manakins are famous for their paired displays, wherein two males work together to attract females.[19] The two males form a stable association that can last for years, though the alpha male is typically older than the beta male. They do not compete with each other, because their roles as alpha and beta are established right from the beginning and do not change, at least not for that pair.

Together, alpha and beta put on quite a show, beginning before any female is present and getting more intense should one show up. These displays take place on arched vines or saplings only about 25 inches (64 centimeters) off the ground, so they are easy to watch. One of DuVal's first tasks was to figure out how to describe the male birds' complex displays, which begin with what she calls a "pip flight," in which the males circle the display court, landing periodically. On landing, the alpha gives the pip call, and then the beta gives a whistle. If a female responds, the males head for the display perch. As things escalate, the two males move from a pip flight to a labored "slow flight" around the area, then progress to the next steps in the display.

Once a female arrives on the perch, the males begin their best-known display, the leapfrog dance. One male hops up, hovers, and lands behind his original perch, while the other male moves to where the first had been, then himself leaps into the air. They do this over and over, pairing each leap with a raspy call to generate a true song-and-dance performance. Males might hover extremely close to the female, but they do not touch her. She follows the display, ducking and weaving on the perch if she is interested, or grooming and regurgitating seeds if not. This final effect is very striking and more than a little comical. (You do not have to take my word for it. There are videos of this performance on the Cornell Lab bird cam. Just search for "Lance-tailed Manakin display video.")

Males actively display on the breeding arenas from February through June and stay in the territories with their display partners in the nonbreeding season. Even in the off-season, males sometime display and even mate, but the big nesting peak is in April.

These duet displays are how male Lance-tailed Manakins attract females to mate, but only one of them, the alpha, ever gets the prize. DuVal and her team figured this out by using DNA markers to see who fathered the one or two eggs in each nest.[20] With extremely rare exceptions (0.7 percent of 413 offspring), only the alpha males had fathered the chicks.[21]

But just because the beta males were fathering almost no chicks does not mean that females were entirely committed to the alpha males. It turns out that females who mated with an inexperienced male, whether he was alpha or beta, were more likely to philander with a completely different male. In a sample of nearly five hundred nests with two eggs, there were two different fathers for the two eggs in about 14 percent of the nests. This includes the rare occasions when betas fathered chicks. Maybe the few females that did mate with betas regretted their decisions.[22] It makes me wonder what all that dancing is for.

So, what motivates the beta males to stay and help in the lek if they don't get mating opportunities out of it? It is rare for nature to reward such apparent selflessness. One reason might be that they are closely related to each other, perhaps brothers. But DuVal and her team did not find that to be the case.

Another possibility is that something else about being a beta made a young male more likely to become an alpha and mate in the future. Maybe males that put in the time as betas are better set up to eventually graduate to the top spot. Indeed, this is what the DuVal team found. A beta one year became an alpha the next year 15 percent of the time; males that were not betas became alphas only 4 percent of the time in her large sample.

In those cases, the alpha disappeared from the study site naturally. But there is nothing quite like an experiment to be sure what is happening. DuVal removed the alpha from eight territories and then observed. Seven of the eight betas immediately began to act like alphas. But that new behavior did not last. Of the four betas she re-sited the next season, only one was still an alpha male. The others had moved back into the beta role with a new alpha partner. I suppose there are alpha characteristics these betas did not yet have.

This early study gave DuVal an idea of why betas might cooperate, but it did not fully answer the question of where alphas came from. By 2013, DuVal had a much larger dataset consisting of eleven consecutive years of data on fifty-seven alpha males' social status, age, and number of young fathered.[23] The observations on who was doing what totaled an astonishing 9,063 hours. It seems like a lot, but DuVal told me: "Oh, it's *so* much more now. . . . 9,063 seems like such a tiny number these days, as we now do 800 to 1,000 hours of observation each year."

I can hardly imagine the tenacity it took to collect this much data in the hottest time of year, right before the rains relieved the heat in a scrubby Panamanian secondary forest, but DuVal told me she is "darn stubborn"—though only for the right things, like huge datasets. And she is still at it, ever expanding the study, bringing new conceptual and technical approaches to her ever-growing manakin world. DuVal gets lots of applications from potential field assistants. Her website reads:

> *We work 6 days a week, from 7 am–6:30 pm, with occasional data entry and organization later in the evenings. Field conditions are hot (frequently > 95 degrees F), extremely humid, and buggy, with steep terrain and a very real possibility of snake encounters, wasp and bee stings, heat exhaustion, and being caught in tropical downpours. Communication is limited at best, and field assistants are sometimes out of touch with most of the world for 2–3 weeks at a time.*

In short, *heaven* to a field biologist.

As important as alpha-beta cooperation seems for Lance-tailed Manakins, there are some intriguing twists. For instance, though nearly all alphas have betas, being a beta is not a requirement for becoming an alpha. DuVal found that 42 percent of the males that became alphas had never been betas. They became alphas straight from having no territory at all. The twenty-seven former betas that became alphas were betas for an average of two years, but some were betas for as many as five years. These betas became older alphas than alphas that had never been betas, and they produced more young in their first year as alphas as compared to the previously non-territorial alphas. This advantage, however, did not last and might simply be explained by their greater age. Those alphas that had never been betas did not have fewer young over their entire lifetimes, either.

So why are there betas at all? I suppose one could say that being a beta might not have been worse than being out of the mating game altogether, and maybe they survived through difficult times to eventually become the alpha. After all, the comparison available to DuVal was between two kinds of successful alphas: those that had been betas and those that had not. There is another class of males that never had the chance to become alphas. No doubt DuVal's subsequent work will study them.

Thus far, we have explored male behavior that attracts females. But we can also ask what females want in a mate. It is a question that the DuVal team has addressed in multiple ways. They measured mate attractiveness via a male's reproductive success, or the number of babies he fathered in a given year.[24] The biggest result was that the experienced, older males had greater mating success.

It is clear that the females spend a lot of time checking out the males. Young females check out more males, and older females spend longer evaluating a few males.[25] Though any one male clearly has the time to mate with many females, since they invest no parental care in the young, different females choose different males. Some of them even choose the same male across different years, just as monogamous birds do. DuVal found that nearly half of females that could have mated with the same male the following year (because he was still alive) did so.[26] Using the twelve years of data she had at the time, DuVal looked at the seventy-three females that were themselves present at the lek for more than one year along with the male they had mated with the previous year. Of these females, 41 percent mated with a male they had mated with previously.

One surprising discovery is that sometimes lekking birds can bias their young toward one sex or the other. This means that certain mothers will have more young of one sex than the other. We do not know the mechanism, and neither do they, but it clearly happens. This is thought to be an advantage in any species where there is higher variation in male than female reproductive success. Rebecca Sardell and DuVal looked at 968 offspring and found that the probability of those offspring being male was lower when the parents were more closely related.[27] Those more closely related parents (though generally not siblings) also produced smaller eggs.

Leks are among the most complicated mating systems we know. Exactly how they function is a challenge only researchers really dedicated to understanding long-term details of a species take on. Perhaps this is why each of these birds has inspired scientists to use novel techniques: Patricelli and Krakauer have a robotic female. DuVal has lifetime data on individual birds. Lank has the genetic details of a chromosomal inversion. But even with these tools, lekking birds have secrets still worth untangling.

CHAPTER 7

Mate Choice and Parental Care

Competition in the Family

When I watch birds, what I most enjoy is observing what they do together, their social lives. Just yesterday I saw a House Sparrow fly up to another, then drop and flutter its wings, begging for food from what was surely its mother. I could not tell that the bird was young from its appearance, but its behavior told me otherwise. The mother did not feed the young bird, instead darting into the bushes, her offspring in hot pursuit. I guess the little one was supposed to find its own food by this age!

Interestingly, it is not just the young birds that get fed. In the willow outside my study, I saw a male Northern Cardinal tenderly offer a seed from our feeder to a female sitting next to him. She could have flown to the feeder herself, but the male seemed intent on demonstrating his ability to care for her. I mused about whether this was why human brides and grooms feed each other morsels of wedding cake.

Usually both parents share in parenting duties. It takes work to get babies to independence, and the nest is a dangerous place to be in the meantime. Both parents are usually needed (and neither is likely to want their brood to fail), yet this doesn't mean they share the labor equally. Each will want the other to do more work.

Family competition extends to the young in the nest, who compete with one another for resources. Indeed, the ways bird parents and offspring resolve their conflicts and maintain their mutual interests make for some great stories. But there is always a background of cooperation that gets the babies reared into successful adults. This chapter covers mate choice, parental care, and competition among the members of a family. These topics are hard to tease apart because mate choice is influenced greatly by who will be the best parent.

Before we get into complex social lives, it makes sense to

start with simple ones: birds where only the fathers care for the young, something that goes way back to the origins of birds themselves. By now most people have heard that birds are the only living descendants of dinosaurs. More specifically, birds are the descendants of a group of theropod dinosaurs from the Jurassic Era (about 150 million years ago). This bird-dinosaur connection was made after the discovery of a fossil that looked like a dinosaur . . . but had feathers. Called *Archaeopteryx*, this fossil was found in Bavaria in 1875 and provided timely evidence for the intermediate forms between fossils and modern organisms posited by Darwin's theory of evolution by natural selection, at least for this dinosaur-bird case. No wonder the Natural History Museum Berlin has a special chamber for *Archaeopteryx*! When I visited Germany, I made sure to see the most famous original in its dark shrine.

Since 1875, researchers have found many fossils on the lineage that led to modern birds. My favorite is one in the Gobi Desert in Mongolia told to me by my friend Amy Davidson, then a fossil conservator at the American Museum of Natural History. It is an oviraptor sheltering a clutch of about twenty eggs, smothered in a long-ago sandstorm and, like Pompeii, immobilized forever in place. Amy told me that a colleague of hers, Luis Chiappe, had spotted a piece of the fossil sticking out of the soft sandstone very close to their Gobi camp. The researchers began by isolating the oviraptor, trying to minimize the work. They did not want to dig any more than they had to, because everything needed to be wrapped in plaster of paris to preserve and solidify it, but the water necessary to mix the plaster was in very short supply. Water was so scarce, in fact, that the researchers did not bathe in the six or so weeks they spent on the expedition. Amy told the story:

> *The Gobi sands are orange and soft and the bone is ivory white. As we followed the exposed claws and forelimb into the sand and cleared it we realized that much of the skeleton was present, and we could see it was in a crouched, brooding position—which was unexpected and intriguing. Luis determined that it would be important to take out the entire skeleton, but we had a very limited supply of plaster—shipped all the way from New York as it was not available in Mongolia. To make the smallest jacket possible we had to try to cut the skeleton out of the rock as closely as possible without damage. As I was reducing the sandy area inside the flexed forelimb and claws with my rock hammer, I heard an odd crunch and saw a greyish fragment had popped out. Luis took one look at it, recognized the bumpy patterns on the surface as eggshell, and at that moment, we knew that the skeleton was brooding on top of a nest—unheard of. I had an odd sensation of empathy for this animal as a female protecting her eggs, but now I know it could have been a male.*

Today, this touching exhibit shows the bones, the eggs, and the plaster shell that encompassed all.

Is the parent huddled over those fossil eggs a mother or a father? David Varricchio and his team at Montana State University looked into this question using a number of fossilized clutches and found that in general, the dinosaurs huddled over eggs are fathers, not mothers.[1] Further, the clutches seemed to be too large to have been produced by any one female. The *Troodon* and Oviraptoridae fossil nests had as many as twenty-two to thirty eggs.

Another kind of evidence that shows we are looking at

fathers and not mothers comes from their bones. Female birds and female dinosaurs that are about to lay eggs store calcium and phosphorus in a spongy area inside their bones called medullary bone tissue. (This tissue is mostly resorbed during egg laying.) That the bones of the dinosaurs associated with nests did not have medullary tissue indicates they were males. Looking at contemporary birds, Varricchio and his team found that in birds with a similar body size and clutch size to the oviraptors, such as ostriches, fathers usually looked after the young of multiple females.

We can only imagine the kinds of interactions between those paternal dinosaurs and the females whose eggs they brooded. Perhaps the females fought over a male, mating with him and then laying eggs in his nest. Or perhaps he had so many eggs because he accepted any female's eggs, and females had to judge who would best care for their eggs.

With the extinction of the non-avian dinosaurs, studying the species that make up the order Palaeognathae, the most basal group of birds on the tree of life, gives us our best guess at how the earliest birds might have parented. Palaeognathae includes many of the bird species that might be easily mistaken for their extinct dinosaur cousins; the group includes the large flightless birds, like ostriches, rheas, emus, and cassowaries, as well as the bizarre kiwis of New Zealand and the tinamous of the neotropics. In some modern species, just as with the fossil *Troodon* and *Oviraptor*, only the male takes care of the young.

This is the case for Greater Rheas, which occur mostly in Brazil and Argentina.[2] Rheas are large, averaging about 50 pounds (23 kilograms). They are long-legged birds with diminutive wings and shaggy, hairlike feathers. Nearly all their lives, Greater Rheas live in groups. Males, females, and juveniles spend the nonbreeding season in flocks. As breeding season approaches in November and December, a male rhea becomes aggressive and courts a group of females. Those he wins over will ultimately follow him to flood-resistant higher ground, where he will trim the grasses and create a small mound built up with soil and sticks and lined with grasses and feathers—the perfect place for a female to lay eggs. Female rheas do not necessarily stay with one male; they may visit several males and their nests, mating with each and laying an egg or two in each nest.

Gustavo Fernández and Juan Reboreda studied the nesting success of Greater Rheas in the pampa grasslands near Buenos Aires.[3] Most nests ultimately held twenty to forty eggs. Nests gained about two eggs a day, which, because any given female can produce only one egg every two to four days, means that at least four females were laying in each nest. Once the young hatched, they scurried about, herded by their worried fathers. Males usually shepherded about fifteen chicks each. By the time the chicks were about six months old, they left the male and joined winter flocks.

Males do not breed every year—ultimately only about 5 percent of male rheas in this population bred successfully each year. This is because a male needs to be in excellent condition, with sufficient fat reserves, to incubate the eggs for around forty days, then herd and defend the chicks for another few months.

I was surprised to learn that, in birds, male care of the young might have evolved before female care. I guess the relationship between a mother and the egg she just laid is not the only one that matters. After all, she has already expended a great deal of energy in laying eggs that the males have not, so they may be better equipped energetically to care for the young. Of course we have only a little evidence of parental care in theropod dinosaurs, and it might not have been the only mating system in those times long ago. A lot of other kinds of breeding systems have evolved.

To my knowledge, there is only one group of birds in which the young are not cared for by *any* adult. In the Australian Brushturkey and its relatives in the family Megapodiidae,[4] males tend a mound of fermenting vegetation that will eventually bring the eggs to incubation temperature. I saw an Australian Brushturkey in Cairns, Australia, and then an Orange-footed Megapode on a field trip farther north with bird guide Del Richards. We saw the megapode in the forest tending its large mound with vigorous backward swipes of one foot, then the other. Orange-footed Megapodes might use the same mound for as long as forty years.[5]

In the megapodes, a female mates with a male, lays a large egg deep in the mound, and leaves. Megapode eggs never experience the warm touch of an avian belly; instead, they are kept warm by fermenting leaves, carefully adjusted by the male. Sometimes multiple females lay eggs in the same nest, or a single female will stick around long enough to lay more than once.[6] This means that any one mound may have numerous eggs that hatch at different times. When each egg hatches, its former occupant claws its way out of the egg, then pushes up through the fermenting leaves to breathe fresh air. Once free, the youngster runs off into the bush, where it will feed on leaves, buds, seeds, and invertebrates, never having met mother, father, or any nestmates. There's none of the bickering, negotiating, cooperating, and cheating that might go on in the other bird families we will meet.

Male care may be the oldest known form of parental care in birds and their antecedents, but now it is unusual.

Andrew Cockburn investigated the literature on parental care in five thousand bird species and discovered that over 80 percent of them exhibit biparental care, while 8 percent have female-only care, 1 percent have male-only care, and 9 percent have more than two adults caring for young.[7]

Why so much care from multiple adults? It is hard to say. Some think it's the natural result of the energy needs of warm-blooded young that hatch from eggs. Perhaps the magic of flight makes biparental care advantageous. Birds travel a tightrope between having enough energy stored and maintaining the svelte figure necessary for liftoff and flight. Besides keeping a trim physique, birds have many specialized features that make flight possible. Feathers, for instance. And their four-chambered lungs, which extract more oxygen from the air than our two-chambered ones.

Baby birds may often need both parents to care for them, but each parent has its own interests. Here we will first explore the mating system and then how parents divide up care. Two parents may care for the majority of birds, but that does not mean they are monogamous in the sense we expect of our spouses. Birds that are not socially monogamous can be polygynous, with several females nesting on one male's territory, like Great-tailed Grackles do; or polyandrous, with females leaving eggs with several male caretakers, like Wattled Jacanas.[8]

One of the big surprises in ornithology has been how often social partners do not match the genetic parentage of the young they care for. This discovery started with a study on Red-winged Blackbirds.[9] The goal of the study was population control, since these birds eat grain.

In 1971 and 1972, working in Colorado, Olin Bray and his team first tried to tamp down the population by vasectomizing male Red-winged Blackbirds. Red-winged Blackbirds are polygynous—males can have as many as four mates—and so it was thought that a single sterilized male would effectively remove up to four nests' worth of eggs from the next generation. The first year, though, the females nesting in the territories of vasectomized males nonetheless laid mostly fertile eggs. Since females were unlikely to know their males were not producing sperm, the researchers concluded that mating with males who weren't their partners was a normal part of the blackbirds' behavior. This astonished the community of scientists, many of whom had considered birds to be faithful and industrious partners and good examples for the rest of us. But it got even more interesting.

Since Bray's discovery, techniques for measuring parentage have improved, and we can now use DNA testing. A couple decades after Bray's work, Patrick Weatherhead and Peter Boag looked in more detail at how female Red-winged Blackbirds chose their mates, both within and outside the social bond.[10] They figured out paternity for an astonishing 617 nestlings over six years. Their data showed that the largest males attracted the most mates, regardless of the quality of their territories. However, just because a large male had a lot of nests in his territory did not mean he was the father of all the eggs. Older, more experienced males were chosen more often than younger males for dalliances outside the pair bond. Weatherhead and Boag called this practice cuckoldry, and they noted that the cuckolders (the "other men") were older and lived longer but were not larger than the females' social mates.

Red-winged Blackbirds were just the beginning. More than forty years after the discovery of extra-pair young in

Red-winged Blackbirds, Lyanne Brouwer and Simon Griffith were able to compile more than five hundred studies of 342 species of birds to see to what extent the young in the nest could not have been sired by their social father.[11] Extra-pair offspring occurred in three-quarters of all bird species studied. In socially monogamous species with biparental care, an astonishing one-third of all broods contained young that could not have been produced by the male tending them.

Why does a bird that has teamed up with another to do the hard work of raising young so often mate outside the pair bond—in a human sense, betraying their partner? Some of the costs of mating outside the social bond are similar for males and females: the cost of searching for another mate, the costs of being away from the nest, and possibly the cost of catching a disease. If the female's mate detects her infidelity and realizes the eggs in the nest are not his, he may care less for the young, a cost only a female is likely to pay. On the other hand, if he is unfaithful, those eggs are elsewhere, and the female he is mated with will still have her own eggs. Basically, there is an asymmetry in which a female's infidelity is more costly to her male partner than a male's infidelity is to his female partner.

The advantage to males seems clear: If they mate outside the pair bond, they can have more young, and others will bear the cost of raising them. The advantage to females is less clear: An unfaithful female will not get any additional young. So why mate with multiple males? She might do so as fertility insurance, in case her social mate is infertile, or because a different mate might be genetically superior to her social mate. She might also get foraging rights or other direct advantages from the neighbor, since females typically philander with neighbors.

Who do females choose to mate outside the pair bond? We can start by looking at a lovely set of studies on Indigo Buntings. Indigo Buntings were my "spark bird": the bird that really got me to pay attention to birds. Long ago, at home in Michigan, I saw an Indigo Bunting sitting high on a dead snag as I looked through my father's heavy old German binoculars. It was on the edge of a meadow near what were then community gardens in East Lansing. This apparently dull bird had a lovely double-noted song. Suddenly it moved, and I was shocked as its brilliant blues gleamed in the sun. It was actually not dull at all. It made me want to dedicate myself to birds.

In a way, Indigo Buntings were the perfect spark bird for me. These are birds of edge habitats, of second growth, uncommon in both prairies and forests but common in the borders between the two. Those borders were the habitat of my childhood. I remember as a teen the disappointment I felt when I learned that the open meadows and brushy thickets that felt wild to me were actually second growth after the felling of nearly all of Michigan's ancient trees.

Indigo Buntings, with their short, fast lives, are one of the "grasshoppers" of the bird world, standing in great contrast to the seabirds of chapter 5. After a winter in the northern neotropics, the male buntings fly north and set up territories. Each female chooses a male and builds a low nest in his territory. If she loses her first nest, she will build another, even in vegetation as ephemeral as goldenrod stems.[12] The males are attentive to their females, copulate often during the egg-laying period, and are diligent about warning if a threat is near.

A typical Indigo Bunting nest will have three eggs, which are incubated by the female for twelve or thirteen days. When the young hatch, mostly all on the same day, the mother eats the eggshells. The mother feeds the young and is occasionally helped by the male, though he is more likely to help once the birds are fledged. The young leave the nest nine to twelve days after they hatch. The parents feed them for about three more weeks, sometimes splitting up the brood and caring for them separately. Then the parents might nest again.

To really understand the breeding system of this lovely bird, I turned to the careful work of David Westneat, whom I have admired for decades. Westneat studied a population of Indigo Buntings in Michigan. He banded and watched the birds on about a dozen territories over three years.[13] During that time, he observed a remarkable 413 copulations. About 13 percent of males' mating attempts were with females on another male's territory (sometimes that male chased the intruder away). Females appeared to resist most mating attempts by intruders, but the genetic data say they were not particularly successful in this resistance.

Westneat compared his observations of rogue matings with the genetic parentage of the young in the population. Over the three years of his study, he sampled 98 families and their combined 257 young. To match young to their genetic parents, he used a technique based on protein variation. He found that about 15 percent of the young were fathered by a male that was not the social parent. This is roughly the same as the percentage of mating attempts from territorial intruders that he had observed. Later work using DNA techniques indicated that, actually, about 40 percent of the bunting young came from non-caregiver males. Though the social fathers did defend the nest, it's no wonder they did not feed the young very often.

Females were more likely to mate with a neighboring male when they were paired with a one-year-old male and the neighboring male was older. It seems that their age might be an indication of genetic superiority, demonstrating their survivability.

Another bird, from the other side of the world, adds some interesting insight into the quality of males chosen for extra-pair dalliances. Great Reed Warblers commonly breed through much of Eurasia and winter in sub-Saharan Africa.[14] They are one of my favorites from my time in Berlin, when I first saw a tan and brown bird perched awkwardly on a slanting reed. It was not the bird's form that most caught my attention that bright May morning as I came down to the second beach on Grunewaldsee. It was its song, if the improbable squawks, cheeps, chirps, and rasps coming from this small bird could be called a song. I knew the vocalization was distinctive, but my *Collins Bird Guide* is not arranged by song, so I had to hunt a bit to identify it. The bird was large for a warbler, didn't seem to be a flycatcher, and certainly was not a thrush or a tree creeper. I went with warbler and, using the guide, quickly came upon its true identity: Great Reed Warbler. The size, habitat, and location matched, but what clinched it was that song, easily played from the *Collins Bird Guide* I had on my mobile, once I knew which one to try.

The tune that first brought me to this lovely bird was what's called the long song, made up of ten to twenty syllables, both repeated and variable. We know this partly from the work of Clive Catchpole, who studied these startling songs in southern Germany.[15] One example of a long song starts with two syllables separated by about a quarter second, followed by about ten additional paired notes in three different patterns. Long songs are thought to attract females, while short songs (usually only four syllables, with two repetitions of two patterns—for example *squack squack, squink squink*) are a prelude to mating.

I cannot resist telling another story about singing Great Reed Warblers. Their songs have evolved to attract females, but they also have a sinister eavesdropper, someone who wishes the males every success in attracting females to their territory, but whose interests are quite different.

This eavesdropper is the Common Cuckoo. Cuckoos cannot raise their own young, but instead lay their eggs in the nests of others. Those cuckoo babies hatch more quickly than their adoptive siblings, then promptly begin tossing the rightful eggs over the nest's edge, one after the other, until they are the sole occupants. It is a stunning sight to see a small Great Reed Warbler feeding a huge cuckoo chick. The cuckoos aren't indiscriminate; their lineages specialize on certain host birds. The lineage that specializes on Great Reed Warblers is very successful, often locating the territory of the male warblers simply by listening for their calls.

I could go on with other stories of cuckoos and Great Reed Warblers, but there is more to the song story.

Catchpole found that female warblers preferred males with fancy songs. Males that attracted no mates had an average repertoire of about sixteen motifs, those with one mate had an average repertoire of eighteen motifs, and those with two females nesting in their territories boasted twenty different motifs to their songs.[16] It makes me want to go back to Grunewaldsee to count the motifs in the males' songs! Besides attracting more females, males with more variable syllables in their songs also produced more young, at least partly because they had better territories with longer waterfront edges. More song and a larger territory seem to be attributes of stronger males.

A female chooses both a territory and the male that possesses it, so it is hard to disentangle the importance of male quality. Since she mostly only gets genetic material from the males she mates with outside the pair bond, she should choose a male of the highest quality.

Determining whether this was true was the goal of a study in Sweden at the far northern edge of the range of Great Reed Warblers.[17] There, Dennis Hasselquist, Staffan Bensch, and Torbjörn von Schantz banded and genotyped the warblers so that they could determine who fathered the young. In all cases a neighboring male fathered the young not attributable to her social mate, but not just *any* neighboring male—the females chose the males with more variable syllables in their songs.[18] In no case did the females mate with a neighbor with a smaller repertoire than their own mate.[19]

It makes sense that females should mate outside the pair bond when neighbors have larger repertoires, because it turns out that the young birds' survival rates track with their father's song repertoire. Even within a nest of chicks, those chicks fathered by a neighbor with a better song repertoire were more likely to survive than nestmates from the social parent. They must have been higher in the genetic lottery for good survival genes. By visiting the Caruso next door, females increase the odds that their chicks will survive to adulthood.

There are other examples of birds with multiple females nesting in one male's territory. Great-tailed Grackles, for instance, nested in the live oaks and pines right across from the biology building at Rice University, where my friend and colleague Kris Johnson studied them and I watched from my office.

Not many people in Houston love the Great-tailed Grackles. They make noises that sound, variously, like the squealing brakes of a train or failing machinery. In winter, grackles gather by the hundreds, often in the parking lot of grocery stores like H-E-B, Central Market, or, in my neighborhood, Fiesta. But the proximity to campus of these usually unloved birds meant that undergraduates could get a taste of real ornithological field research right within sight of the student union.

If you have not made the acquaintance of Great-tailed Grackles, give it time. Their population growth in the United States since the 1880s has been remarkable and related to the growth of human habitat.[20] They were once restricted to southern Texas, south of the Nueces River, but the species reached Houston in the 1930s—right about the time my German Jewish immigrant father's family did, having escaped Germany just in time. The grackles then spread north and west through the Great Plains, extending their breeding ranges into Southern California and up into southwestern Minnesota by 2000. Now Great-tailed Grackle nests commonly adorn trees planted in the scraps of land created by freeway entrances and exits in Texas and beyond.

Male grackles spend much of their time together on the ground. I often see them sidling up to one another, then pointing their beaks to the sky, the better to assess their relative size. Even the young first-year males, with little chance of attracting a female, contest one another harmlessly on the ground, uttering their unwelcome squawks. The real action is in the nesting trees, though, where females weave nests of weeds and grass. A smaller tree might have a single male, while larger trees may have several.

Kris Johnson and her team looked at the family life of this noisy species. They banded the adults by coaxing them into a large walk-in trap, while they banded the young at about twelve days of age, reaching the nests with a cherry-picker lift (operated by Juan Alejandro, who loved to take a break from his regular job to help the researchers). After the banding, the team took small samples of blood from the birds to genotype them, determining parentage.

A floppy hat may seem a useless accessory when studying Great-tailed Grackles on a Texas university campus, but Johnson tried to fool the grackles with one so they would not attack her while she did her daily patrols. The hat didn't work, but limping did. The grackles seemed to identify Johnson more by her gait than by her attire and could not recognize her when she limped. Why were they attacking? Well, on these patrols she counted nests, searched for banded females and males, and determined where one male's territory ended and the next began. Males in particular would bombard her repeatedly because they recognized her as the intruder who rose high into the tree and plucked babies right out of their nests. A grackle's brain is not wired to understand that Johnson also returned those babies to their nests, now ornamented with leg bands and missing less than a drop of blood, so she depended on her disguise. It allowed her to observe the birds and see how vigorously the territorial males defended their nests from potential predators with alarm calls and mobbing. They were, Johnson said, very good at the defense that is such an important part of paternal care: "A couple of angry male grackles could be quite effective at deterring avian or mammalian nest predators."

Behavioral observations told Johnson which bird did what, but only the precious blood samples could tell her that a single male, banded with three bands, of green, red, and gold, sired 40 percent of the 120 genotyped babies, 14 in nests others defended as their own. Though this popular male did not father all the young in his own territory, he was in a better situation than the male with three blue bands, who diligently defended a nest of chicks in a lone pine tree, all of them the genetic offspring of the green-, red-, and gold-banded male.

Overall, 37 percent of nestlings were sired by a male other than the mother's social partner. Of these, 13 percent were sired by resident bachelors that did not even have their own territory. In Great-tailed Grackles, females exhibit quite a bit of choice.[21] They choose to nest in territories of heavier males with longer tails, and they seek the same attributes in their mates outside the pair bond.

Other birds with even more active social lives are worth exploring. Superb Fairywrens alone might be worth a trip to Australia. These tiny little sprites flutter about the underbrush, flashing their brilliant blues and turquoises and confusing anyone interested in monogamy. Andrew Cockburn and his team at the Australian National University have spent decades studying these birds, which typically have more young by males that are not their partners than by their partners.[22]

Radio tracking reveals exactly how fairywrens get together to mate outside the pair bond. Females initiate contact with extra males in the half hour before dawn, usually two or three days before they begin to lay eggs. They leave their social group and fly to where unattached males sing in the dawn chorus. There they pick males with particularly long trill songs, often males they recognize from the months the males had spent courting them with song and color.[23] Males sometimes augment their color by carrying a yellow flower or petal as they display. Those that molt the earliest, and thus begin to display to females earliest, get the most matings even when mating does not take place for months.[24] This preference for early molting is one of the few examples of a trait females prefer other than age.

When the female returns to the nest after her philandering foray, her mate joins her and attacks any other male who comes near. The female sits on the nest a bit, then flies to a branch and solicits her mate. The ensuing copulation is longer than usual. At this point, the female will have sperm from two males battling it out in her reproductive tract.

Amazingly, Sara Calhim and her team found that there were two different forms of sperm: One, sperm with short flagella and large heads, was better at fertilizing eggs from extra-pair matings; and the other, sperm with long tails and small heads, was better at fertilizing eggs from pair mates.[25] Females mated with males who were not their social partners, and then when the females returned to their territory, they mated with their social partners. Their mate's sperm could swim faster, displacing some of the previous male's sperm. Any given male made only one kind of sperm according to whether his reproductive success was more likely to come from a social mate (long tails, small heads) or a mate outside the pair bond (short tails, large heads). Clearly, Superb Fairywrens have a complicated social life—for us, it reveals the power of male quality in female choice.

Up to this point we have explored how male and female birds join together in pairs or more, with or without fidelity. The next question is how do males and females divide up caregiving at the nest? Just because mates are chosen does not mean that there is no more strife. It is common even in apparently monogamous birds for females to provide more care than males do.[26] After all, their interests are not identical. Some of the young may have been fathered by another male, reducing the tending male's genetic interest in the brood. This is something he may or may not cognitively realize. What is important is how he behaves, and that will come from both learned and evolved behavior. The male may put energy into finding other females to mate with outside the pair bond so that he can have progeny he does not have to care for. Survival to the next season is important and a separate interest for each, especially if the male and female are unlikely to pair together the next season.

The exact extent to which males and females invest in their brood is not necessarily fixed. If a female thinks she is paired with an exceptionally high-quality male, she might invest more in that particular brood, believing they will have a higher chance of success. The same can be true of the male. Nancy Burley calls this the "differential-allocation hypothesis."[27]

When I knew Nancy Burley, we were both graduate students in zoology at the University of Texas at Austin. I knew her better than many students because we both worked at the Brackenridge Field Laboratory. With her large outdoor enclosures for pigeons, Burley did experiments on mate choice. I was following wild nests of the paper wasp *Polistes exclamans*, some of which nested on the pigeon cages or on the empty cages where Burley conducted experiments. She had a couple years more experience than I had and was full of good advice—in particular, about the power of a careful experiment. It was welcome advice; at the time I was mostly just marking the wasps and watching to see what they did. This was the mid-1970s, so long ago we did not even have Walkman radios yet, so conversation on long, hot field days was especially welcome.

Burley first demonstrated differential allocation in monogamous Zebra Finches, an Australian bird commonly used in lab observations of behavior. She manipulated the attractiveness of males and females by tagging them with colorful leg bands. Learning that female Zebra Finches go crazy for males with red leg bands but consider green leg bands unattractive, and that males are most attracted to females with black, not blue, bands, Burley could change each bird's jewelry, then follow its reproductive success. As Burley predicted, males that mated with attractive partners put greater energy into rearing the young, resulting in higher reproductive success.

The study caused a lot of consternation in the community because most researchers did not want the colored leg bands they used to identify birds to change their attractiveness. It turns out they did not have to worry; the effect seems to be particular to the Zebra Finch as Burley studied it.[28]

Other birds have been studied to see what happens when parents judge their young to be of higher quality. One of these is the Blue Tit, among my favorite birds of the Berlin winter. Their markings seem more elegant than those of the closely related Great Tits, though I like them a lot, too.

Ben Sheldon and a Swedish team looked into whether female Blue Tits that judged their mate to be superior would produce more male chicks.[29] They designed their study because, in natural populations, broods of males with more ultraviolet plumage, which is visible to birds, produce more sons than daughters. The likely cause of this is that those bright males would have higher breeding success than duller males and so producing more males would be advantageous. We do not know the mechanism for how they do this, but it is likely to be influenced by hormones in the female's body at the time the egg is formed. Females have less variable breeding success no matter how healthy they are because a given female can lay only so many eggs. Males, on the other hand, can produce sperm for extra matings at little cost. So, if you are going to have fabulous-looking progeny, better make them sons, at least if you are a Blue Tit.

Sheldon and his team wondered if they could get females to produce fewer sons by masking the males' ultraviolet coloring. Indeed, they found that males whose bright ultraviolet was masked no longer produced more sons than daughters. They do not know the mechanism for changing the sex ratio of the young in a brood but determined that it had to happen before the eggs were laid.

Another study on Blue Tits and parental effort investigated relative male and female effort.[30] In 2003, Arild Johnsen and his team worked on Blue Tits in Kolbeterberg, Austria, near Vienna. Ninety-six of the 250 nest boxes they placed in their study site attracted Blue Tits. The researchers captured the parents and recorded their age, weight, and size. Then they put a tiny transponder on one leg that could be detected by antennae set up on the nesting boxes. They also measured the birds' plumage color and manipulated the males' crown color with marker pens, making some more attractive (that is, more visible in the ultraviolet range) and some less attractive (by blocking their ultraviolet color). The results showed that yearling females mated to brighter males fed more to each chick, though there was no such pattern among older females. The brighter males fed their chicks less than the duller males did. They also found that females mated to brighter males were more active in nest defense than females mated to duller males. It seems obvious that the females knew they were mated to brighter males, but the males must also have realized that about themselves, given they changed their own behavior after the investigators marked them. Johnsen and his team thought that the males realized their status from the way the females responded to them, then acted accordingly—just as a human teenager might do in social interactions.

If ever there was a bird with a flamboyant male, it would be what I call the peacock, though its official name is the Indian Peafowl (the male is a peacock and the female a peahen). The males carry long trains, the group of feathers that they can erect into a beautiful fan covered in large spots. They display in a kind of exploded lek, in which males hold their own tiny display courts in hearing distance of other males. They offer no paternal care. When a female is present, a male lifts his feathers and spreads them into a quivering fan, shaking every feather so the eyespots shine, flashing green, blue, and turquoise. Then he slowly pivots for the female, who most often chooses the male with the most eyespots.[31] This answered Charles Darwin's famous dilemma expressed in a letter to Asa Gray in 1860: "The sight of a Peacock's train whenever I gaze at it makes me sick."[32] The dilemma was that males invested in a trait that did not increase their survival and could even decrease it, since the display feathers also made males more visible to predators. If natural selection governed the evolution of traits, then they should increase, not decrease, survivorship. But what should really increase is the genetic representation of the male in the next generation, and the peacock could do that by mating with many females even if he didn't survive as well.

Marion Petrie followed eleven peahens in captivity to count the eyespots of the males visited by the peahens.[33] For example, Female 1 visited Male A (147 eyespots), then Male B (161 eyespots), and mated with B. Female 8, meanwhile, visited Male A (147 spots), Male F (141 spots), Male D (152 spots), and Male E (157 spots); she mated with Male E. The females always chose the males with the most spots. Males with more spots were not only more likely to be chosen but also had higher survivorship, and more of their young survived. It was clear: The genetic variation among the males was meaningful for females and consequential for both.

Petrie is retired now, but I have met her at a number of International Society for Behavioral Ecology meetings. She is the author of one of my favorite papers of all time, a piece on female choice in moorhens, a bird now called the Common Gallinule. The paper's title almost says it all: "Female Moorhens Compete for Small Fat Males."[34] The male moorhens take on most of the egg-incubation chores, and they lose weight during the breeding season, so the plumper the male, the better he will be at incubation.

A nonexperimental approach to looking at whether males invest more when they think they are mated to high-quality females has yielded fascinating, widespread observations. Female birds often lay blue or blue-green eggs, a coloring that comes from a chemical called biliverdin that mothers deposit into the shells of the eggs. Biliverdin is a potent antioxidant, and it is costly for the birds to produce, so blue eggs are a sign of a high-quality female.[35] Daniel Hanley and his team wanted to look at egg color and paternal behavior. They studied one of my favorite common birds in

St. Louis, the Gray Catbird. In this bird, males establish territories and sing. Females join the males and build the nest. Early on, the female incubates the eggs, and the males bring her food, increasing food delivery when the young hatch.

First, to relate egg color to female condition, Hanley and his team caught the birds in mist nets, measured and weighed them, and took a tiny amount of blood to measure antioxidant capacity. Then the investigators measured egg color with a spectrophotometer.

Male brood care was harder to measure. The investigators set up video cameras and recorded nest visits during which the male fed the young. They found a strong, almost linear relationship between egg color and paternal effort. At the low end, the chicks that emerged from the dullest colored eggs received about one paternal visit per chick per hour, while on the high end, the chicks from the bluest eggs were visited nearly four times per hour.

Just looking at the Gray Catbirds in Ruth Park near my home in St. Louis, I would never guess that the males were so attentive to egg color. I suppose males with females laying duller eggs save their energy for a second brood (as they often have two broods a year), for producing the next year's brood, or for mating outside the pair bond. In fact, in a study of 455 Gray Catbird nestlings in 165 nests, 13 percent of the nestlings (occurring in a quarter of all broods) were from extra-pair fathers.[36]

Juan Soler and his team took a more global view of egg color, hypothesizing that polygynous birds, wherein females strive to get paternal care, would tempt males with bluer eggs if they could.[37] In monogamous species, by contrast, they suspected the females could get away with duller eggs, conserving their biliverdin for other purposes. It is unlikely that the females consciously produced bluer eggs with better males; there was probably some evolved mechanism that resulted in this differential channeling of biliverdin. To test this idea, Soler cataloged the egg color and breeding system of 152 species of songbirds using measurements of egg color from photographs and from 5,878 eggs from 98 species that were held in museums. The results supported their hypothesis: Females from polygynous species lay bluer eggs.

Now I'll turn to the babies—to the ways they compete for and seek parental care. Just about the last place you might expect conflict is between parents and their young. After all, the babies are the parents' hope for the next generation. They are how parents pass on their genes to perpetuity, one generation at a time. Baby birds often come into the world more or less helpless; they would perish without their parents' care. But parents often have more than one baby (and can anticipate having more young in the future) and only so much energy. Further, where you might think chicks would simply let the parents know how much food they need and the parents would provide it, any given chick is likely to want more than just enough food to help it survive and prosper. In contrast, the parent has the most surviving offspring by providing care more equally to all the young in the nest.

Exactly how conflict and cooperation play out depends on kin selection, as laid out by Bill Hamilton in 1964.[38] Kin selection simply means that an individual can pass on genes to the next generation by helping related individuals that are not progeny. These relations could be siblings or cousins, for example. Kin selection therefore favors cooperation at levels beyond the nuclear family.

I met Hamilton as a graduate student in Austin. He

impressed me with his modest brilliance and fascination with the social wasps I was studying. I subsequently had many chances to talk with him while he was a professor at the University of Michigan. I have fond memories of our families spending time together at the University of Michigan's Biological Station, where we went on a walk, talked about bark beetles, and watched our young daughters enthusiastically gather up little twigs for a project known only to them. This was long before Hamilton received seemingly every accolade possible, including the Crafoord Prize, the equivalent of a Nobel in evolutionary biology.

Hamilton's famous theory depends on genetic relatedness among family members and on the benefits of cooperation. It is probably best articulated in Richard Dawkins's famous book *The Selfish Gene*.[39] The theory explains cooperation in organisms like social wasps, bees, ants, termites, and others where there are sterile individuals that do not have offspring to rear and instead rear sisters and brothers, most typically. It can also explain cooperation among organisms that don't give up reproduction to help relatives, for example when individuals help nieces and nephews in addition to their own young.

It was Robert Trivers's insight that kin selection theory could also be applied to conflict within families.[40] I met Trivers around 1974 when I was in my first year of graduate school at Texas and he came for a visit. His paper on parent–offspring conflict had just come out and was receiving tons of attention. We graduate students had read all of Trivers's papers and were awed when he strode in, tall, elegant, and dressed in a long coat, unusual for Austin, Texas. He wowed us with what has become fundamental theory in animal behavior (and he, too, later received the Crafoord Prize).

The difference in interests (or the zone of conflict) between parents and offspring or among siblings, Trivers's theory holds, comes from the differences in relatedness. An individual is, by definition, 100 percent related to itself and only 50 percent related to each parent and to each sibling (provided that sibling shared the same mother and father). So if food is limited, and an individual wants more than the parents want to give it (unless it is the only young in the nest), the zone of conflict will arise from the difference between 100 percent and 50 percent. An offspring wants itself to live most of all. But it also wants its sibling (carrying 50 percent of its own genes) to live, hence cooperation among siblings. But if food delivered to the sibling means that the offspring in question would die or be weakened, it does not want to share, hence conflict among siblings.

The disagreement between mother and offspring plays out in a lot of ways, as each youngster tries to get more food than the parent wants to give it. Generally unlikely to physically manipulate their much larger parents, youngsters sometimes resort to psychological manipulations like intense begging (in children, we call these tantrums). It can be challenging for a parent to disentangle real need from an offspring's desire to have more for themselves than for a sibling.

Luckily, begging is not the only way parents can tell which nestling needs the most food. Another sign is the color of the nestlings' mouths, which in some species (including canaries) flush more brightly upon a parent's arrival when there is greater need.[41] Notably, parents do not necessarily favor feeding the weakest, neediest nestling.[42] A lot of evidence indicates they favor stronger young, sensing when some are not going to make it. Better to have one or two survive at the expense of the others than lose the whole brood.

Begging itself has a cost, because it attracts predators to the nest. Susan Leech showed this by creating artificial nests baited with quail eggs, then comparing nests in which she played recordings of the urgent begging calls of tree swallow broods and those in which she did not.[43] She found that nests with begging sounds were attacked by predators first.

In another study, Leech, along with James Briskie and Christopher Naugler, predicted that begging would intensify among less-related young, such as when the mother had mated with multiple males or when there was a parasitic cowbird chick in the nest.[44] Full siblings, sharing 50 percent of their genes, should be more considerate of one another and beg less. The authors recorded begging calls from eleven species of songbirds for which they knew whether the chicks shared a mother and father or just a mother. The birds they studied included Tree Swallows, Eastern Bluebirds, Dunnocks, Indigo Buntings, White-crowned Sparrows, and Lapland Longspurs. As they predicted, nestlings with lower relatedness begged loudest.

Nestlings may beg and compete for food, but there is another, perhaps more serious competition that plays out between parents and young. It concerns the timing of leaving the cozy nest, where the young can wait for food and snuggle together to share body heat—but at a cost to the family. If a predator arrives, it is likely to eat all the babies at once. This is disastrous for both the babies and their parents, which may explain why nestlings fledge very young, when they can barely fly.

The conflict arises because parents and offspring do not necessarily agree on the best time for fledging.[45] Todd Jones and a large team of researchers looked at the survival of young in the nest and of young that have fledged and left the nest across eighteen bird species. They identified an intriguing difference in the interests of the parents and the progeny. For twelve of the species, the young were more likely to die after fledging than while in the nest. Why did they not stay in the nest longer? It turned out that the nest may be safer for any one baby, but at least one youngster out of the entire clutch is more likely to survive if they have fledged, because a predator will be less likely to find all the babies. The differences were large overall. Any one chick was 14 percent more likely to die after leaving the nest. But if you put the survivorship of all the nestlings together, at least one is 14 percent more likely to survive if all of them are no longer in the nest. So from a given chick's perspective, the nest is safer; but from the perspective of the group as a whole, the parents' perspective, getting them out of the nest is better.

The breeding biology of a father, a mother, and their young, as covered in this chapter, is deceptively cooperative. But it is not so simple. Sometimes multiple females mate with one male, and sometimes one male takes care of the young of more than one female. There are often conflicts of interest between the male caring for the chicks and the female that laid the eggs. Additionally, because males and females mate outside the pair bond in many species, males might have progeny in other nests and unrelated young among those they care for in their own nests. When there is not sufficient food, the progeny will compete with one another for each morsel. This is to say, amid all this cooperation, we also find competition at various points in the mating systems of different species. What comes next is even more complex.

CHAPTER 8

Families with Helpers

Older Siblings, Lonely Bachelors, and More

Outside my window, a Common Grackle is offering a morsel of suet to a bird with fluttering wings and a gaping beak. The two birds are nearly the same size, though the brown recipient lacks the blue-black luster and yellow eye of an adult grackle. I see: This is a parent making an offering to a son or daughter, hatched this summer.

But in some species, an adult bird offering food to a youngster cannot be assumed to be its parent. Sometimes the adult who feeds a begging chick is an older brother or sister from the previous year who, rather than strike out on its own, has chosen to stay and help its parents rear the next brood. Or it could be an unrelated male currying favor and hoping to eventually become the breeding male in the group.

The big question about such helping behavior is *why*: Why would an adult help someone else raise young rather than have its own young? After all, natural selection favors actions that increase an individual's reproduction. So what is happening here? The answer depends on what the alternatives to helping are and how the helper is related to the brood. It can be difficult or impossible for a younger bird to breed on its own if territories are unavailable; in that case, the odds may favor the bird eventually inheriting the territory if it sticks around long enough, and perhaps while helping, it passes along its genes by raising relatives (such as younger siblings). As noted in the last chapter, this sort of helping, known as kin selection, is another way of passing genes to the next generation.

In 1969, with kin selection in mind and colored leg bands in tow, Glen Woolfenden embarked on what would become a decades-long study of Florida Scrub-jays at Archbold Biological Station.[1] Woolfenden was soon joined at the Central Florida site by an undergraduate intern named John

Fitzpatrick. Fitzpatrick would ultimately become executive director of Archbold and, most recently, executive director of ornithology's mothership, the Cornell Lab of Ornithology.

I was also studying helpers at the nest shortly after Woolfenden began his work on Florida Scrub-jays, though I was looking at helpers in paper wasp colonies. I think all of us expected that if there were helpers, the nest, whether one of wasps or jays, would produce enough young to make helping pay off.

It is hard to imagine a nicer place to study birds than Archbold Biological Station, though it is a bit warm in summer. Its 5,200 acres (2,100 hectares) of oak scrub are generally human height or shorter, which makes it possible to visually track birds and reach their nests for bird banding and weighing. After years of being given bits of peanuts, the jays at Archbold are also quite tame. They will land on heads and hands, letting people approach them without apparent alarm. This makes it easy for the researchers to be complete in their censuses of the birds. Even birds that disperse outside the study area remain tame and approach curious researchers, perhaps because of their prior experience with people. Imagine how much easier it is to identify these approachable birds by their colored leg bands than it is a skittish warbler, high in the forest trees, their legs obscured by branches!

Woolfenden and Fitzpatrick found that nests with helpers produced more young than nests without helpers did. It was not a lot more, but helping at the nests was a better option for some birds than trying to find a place to breed in the harsh environment in which they lived. In addition to rearing relatives, helpers sometimes inherited the territory, particularly if they were males and their own mother had perished. But these were far from all the details that Woolfenden and Fitzpatrick explored to try to understand helping in Florida Scrub-jays. Just look at the topics of the chapters in their landmark book: the pair bond, helpers, territory, dispersal, reproduction, survivorship, and evolution.[2]

The best way to understand the benefits of helping might be to leave helpers already present with some nesting pairs of birds and remove them from other pairs. This is the only way to be sure the birds start out in similar conditions. Otherwise, one might worry that those birds with helpers were different to begin with from those that did not have helpers at the start—for example, maybe they had more breeding experience or better territories. This is the experiment that researcher Ron Mumme did with the Florida Scrub-jays at Archbold Biological Station.[3]

Mumme told me that he took ornithology with Glen Woolfenden in 1975 at the University of South Florida and the course included a trip to Archbold. That trip—and E. O. Wilson's now-famous but then brand-new book *Sociobiology*[4]—was enough to hook Mumme on cooperative breeding. Woolfenden helped him join Walt Koenig's project on Acorn Woodpeckers (more on them later), but Mumme never forgot the bird that started it all for him, the Florida Scrub-jay. With the encouragement of Woolfenden and Fitzpatrick, and postdoctoral funding from the National Science Foundation, Mumme went to work planning an experiment on the value of the scrub-jays' helpers.

Of course, Mumme was not permitted to remove helpers from the long-term study population, so he chose an adjacent population for his study. Each year, for three years, Mumme and his team identified families of Florida Scrub-jays with helpers. For the experimental group, during the early nest-building and egg-laying stages, from mid-March to early April, the researchers removed all helpers from some of the scrub-jay families, leaving the nests with just the breeding pair to care for the nestlings. They kept the removed birds in captivity, supplying them with ample food until the end of the breeding season, when they were released in their original territories. For a control group, the researchers used an unmanipulated set of families with helpers.

Mumme's team observed the groups and their success at rearing young, checking the nests every two or three days and weighing, measuring, and color-banding the young eleven days after they hatched. They watched each nest to see how often the parents and any helpers delivered food to the young. And most important, they measured the success of each group in two ways: the number of young fledged at three weeks of age and subsequent survival at independence (which occurred at about two months of age).

It was important that Mumme verify that the helpers were indeed aiding relatives. This was not hard to do, though, because Florida Scrub-jays are faithful to their mates. This had been definitively shown in a DNA-based study of 771 babies from 279 nests, which found that only 2 jays were not the progeny of the mother's male partner.[5] Therefore, Mumme and his team simply used genealogies of the banded individuals to infer how they were related. It turned out that most of the helpers were sons helping their parents raise their siblings.

The researchers observed that pairs of breeding adults whose helpers had been removed fledged half the number of young as the adults that kept their helpers. Nestling deaths were mostly due to predation, not starvation. It usually happened during the day, with attacks by other birds and snakes being the most common causes of mortality. Helpers reduced these attacks by warning of approaching danger. Once warned, all the jays in the area mobbed the predator, often prompting its hasty retreat. Even though predation was an important factor, it was also true that young in families with helpers received more food, so they became larger and therefore more likely to survive.

Mumme then asked exactly how much of a difference helpers made. He measured success when the young were two months old and quite independent, calculating that each helper added about half an offspring to the group. This means the family line and the DNA it carries benefited from the help.

It is hard to say what the helper was giving up by helping, since independent breeding might not have been possible. But besides the DNA benefits of contributing to rearing relatives, it is also possible that helpers could receive direct benefits—some advantages that ultimately increased that individual's opportunity to breed. For instance, a male helper could become the breeder and inherit the territory if the female there was not his mother (a female helper typically leaves the territory to breed).

Another question was, what alternatives were available to the helpers? That was taken up in another study of Florida Scrub-jays that looked at how well the birds reproduced in different habitats. Florida Scrub-jays are entirely dependent on the increasingly rare oak-scrub habitat found on relic sand dune ridges scattered in an irregular line from northern Florida down the center of the state to the Everglades. Periodic fires started by lightning strikes keep the scrub at the low and variable level preferred by jays. Along with other human activities, like development and farming, human efforts to suppress fires can actually harm Florida Scrub-jays. After about fifteen years without a fire, the vegetation becomes too high, and the scrub-jays are replaced by Blue Jays.

Florida Scrub-jays are now considered vulnerable by the International Union for Conservation of Nature. This means that the birds' opportunities for reproduction might now be more limited than they were historically. If, as they mature, the juveniles have nowhere to go to nest on their own, perhaps they might as well hang out with mom and dad and occasionally gather food for the young or warn of a predator. But before concluding that, it is best to study how the birds do when nesting in different habitats.

The low oak-scrub habitat is already full of scrub-jays, but neighboring habitats might suit a young bird attempting to leave its family role as helper and breed independently. David Breininger and Geoffrey Carter studied the usefulness to scrub-jays of multiple habitats at Happy Creek, a 1,000-acre (400-hectare) long-term habitat restoration site right on the Kennedy Space Center grounds in Florida, not far from Cape Canaveral, where humans launched to the moon.[6]

For this study, Breininger and Carter sorted the habitat into four categories: one with vegetation at optimal height, about six feet tall with a mix of shorter vegetation; one with shorter vegetation; and two others with taller-than-optimal vegetation. The optimal territories were burned about a decade ago, but along with the low plants, the birds like to have open sandy areas that burned recently in their territories. Too much fire and the vegetation is too low. Too little fire and the trees become too tall. Too much uniform fire coverage and the mosaic of plants as measured by their heights is lost. It is a balance that has been honed across millennia by natural fires caused by lightning.

Breininger and Carter found that all the optimal territories were occupied during the study, so new breeders would be forced to select from territories in which vegetation was too tall or too short. Could they succeed there? Generally, they could not. More birds went *to* the suboptimal territories than were produced *from* them. In other words, the population size increased only when the birds bred in optimal territories. The suboptimal territories are what ecologists call "sinks"—places where birds do not flourish and their population declines.

While the scrub-jay studies are a good beginning for understanding helping, there is a lot more to the story of helpers. Let's go to Australia, home of the Superb Fairywren and the superb scientist Andrew Cockburn, who researches them.[7] Andrew helped organize the 1996 meeting of the International Society for Behavioral Ecology in the country's capital city, Canberra, where I saw his Superb Fairywrens and went to many talks on cooperative birds. He has brought the field forward in many ways, and has a keen eye for detail and a healthy skepticism, which we will hear more about later.

There are lots of helpers in Australia, particularly in eucalyptus forests.[8] Australia is also home to the only avian family in which all species have helpers—Maluridae, the fairywrens.[9] Fairywrens are unrelated to northern hemisphere wrens, but they do have upright tails and are similarly small and active. In fairywrens, the helpers are generally sons of the dominant female. The breeding female is unrelated to her mate and does not mate with her sons.

My own youngest son, Philip, spent an austral spring following Variegated Fairywrens through the scrub near Brisbane, Australia, hunting hard for nests. He was a research assistant to a graduate student from Cornell University. Philip thought he knew the landscape well but told of a night when he climbed a hill to watch the sunset and became disoriented. Unwilling to get even more lost or to climb down the wrong side of the hill, he sheltered against a rock wall for the night, glad to finally hear the birds in the morning and make his way back to the now visible study site.

I love Superb Fairywrens. *Birds of the World* vividly calls them “perky light-footed jewels,” a description that fits so well. Their colors are improbably bright but could also be viewed as camouflage, since their patterns of dark indigo and bright turquoise mimic sun flecks in the understory. When I saw the Superb Fairywrens at the botanic gardens in Canberra, many had colored leg bands from a study that began in 1988.

In Superb Fairywrens, the background color of the adult male in mating season is black. It frames two color patches. The first is a bib patch, which looks dark to humans but is quite reflective in UV light. This patch is probably obvious to other birds when the males throw back their heads and sing. Lighter blue feathers on the males’ crowns, napes, and cheeks are flared in displays to prospective females. They brightly reflect the pale blue, but the structure of the feathers also means that they perfectly reflect white light like a mirror, so-called specular reflection, allowing the male to dazzle the female. Females are also lovely, if more subtle in color. They have blue tails tipped with white, bodies in various shades of brown, white bellies, and orange around the eyes, which matches the orange of their beaks.

Superb Fairywrens stay on their territories all year round, leaving them occasionally to join winter feeding flocks. Females maintain territories, build nests, and pair with the dominant male on the territory. They lay up to four clutches a year, beginning a new brood almost as soon as the previous brood has matured or failed.[10]

If we want to understand how helpers are related to the brood they help, fairywrens do not make it easy for us, because unlike Woolfenden's scrub-jays, breeding fairywrens frequently mate outside the pair bond. For a helper, raising half siblings is—appropriately—*half* as lucrative a strategy as helping to raise full siblings, and so we should expect helpers to have stronger incentives to seek their own breeding opportunities. Nonetheless, helpers are common in this species. They actively feed and defend both nestlings and fledglings.

Though helping may not seem to be as advantageous as breeding, it is often temporary—more than half of Superb Fairywren helpers eventually become breeders on the same territory. Two things must happen first: The helper's mother has to die (thus she is no longer the breeding female on the territory), and the helper has to become the oldest helper on the territory. Helpers queue up to take over, leaving younger birds without a chance. Occasionally younger males will take up with a young female who lacks a territory and lay claim to a piece of her natal territory. Indeed, male fairywrens depend on the presence of females and will move to a nearby territory if there is an unpaired female there.

Females, on the other hand, leave their territories in their first year and look for an opportunity to breed. If they don't find a chance to breed, they join another group as an extra helper. By the next summer, in August, they again actively look for territories in which to breed. By November, they have all dispersed, and none are helpers. There is an advantage to leaving early, because a female is more likely to win the breeding position in a new territory if she has been there awhile.

Cockburn and his team studied Superb Fairywren helpers for two decades to get a clear idea of who the helpers assisted. They found that half the male helpers were helping two birds that were not their parents, a quarter were helping their mother and their father was gone, and the rest were either helping their father and their mother was gone or were helping both parents.

Since helper males are not often related to the young they rear, it is a bit of a surprise that they help at all. Yet they do, at levels that vary considerably. Their care is essential to the success of the brood, but they save enough energy to court females outside the family with visits and song in the dawn chorus. Anne Peters and her team found that there was a balance in testosterone levels in males where both courting and paternal care could occur.[11] When she treated males with more testosterone, they cared less for the young. Males do not adjust their care according to the needs of the young. Instead, Cockburn found that it is the females who make these fine adjustments, increasing their care if the males reduce theirs.

Females also lay smaller eggs with up to a 20 percent lower nutrient content when they have helpers.[12] The smaller young that hatch from these eggs need more food to catch up, but that is provided by the helpers. To study this, Andy Russell, now at the University of Exeter, switched eggs laid with and without helpers to see what would happen; the eggs laid with helpers but reared without them produced lighter fledglings. Cockburn and his team agreed that the cost of egg production should be included in any study of helpers.[13]

Helpers may be most common in Australia, but that does not mean they do not appear elsewhere. Another bird that uses helpers is among the best-loved in Europe: the Long-tailed Tit. The Long-tailed Tit is unrelated to other tits or to what North Americans call chickadees, and the story of its helpers is worth telling because they seem so different from other birds' helpers.

No Long-tailed Tit *wants* to become a helper. They all start out building their own nests as early as February, according to Ben Hatchwell, who has spent his career studying them.[14] And these are not just any old nests. They are domed masterpieces made from mosses bound with spider silk, as if made by fairies, then lined with as many as 2,500 feathers and coated with thousands of tiny flakes of lichen to camouflage the exterior. These nests take more than a month to build, though replacement nests can be hastily built in a dozen or so days. Females lay about ten eggs, and begin to incubate them only on the day the last egg is laid so that they hatch at the same time after a couple weeks. Her mate feeds her while she is incubating, then feeds her and the young for their first week while their mother keeps them warm.

What is heartbreaking about this cozy story is that about three-quarters of all Long-tailed Tit nests are destroyed by predators, mostly crows and jays, but also weasels and stoats. When that happens at the egg stage, the parents build a new nest. If the eggs have already hatched, they do not rebuild. Instead, they either find another pair's nest, where they can help, or they sit out the breeding season.

Fifteen percent of adults end up as helpers, and each nest that is successful has an average of two helpers.

Long-tailed Tits that have lost their nests prefer to help relatives. Three-quarters of helpers pitch in at a nest of a parent or a sibling, though usually not a full sibling, since "divorce" and death rates are high. They might even help at the nest of one of their own offspring, helping their kids from a previous year raise their grandchildren. The helpers do not have an impact on predation rates or on the survival of the nestlings, but they have a big impact on the condition of the young, as measured by young that survive to join the population the next year. Hatchwell found that a male reared by just its parents was only 15 percent likely to survive to join the breeding population, while one reared with four or more helpers was 40 percent likely to survive (these numbers are calculated for males because females are more likely to disperse and so their survival cannot be so easily determined).

Another benefit of helpers is that they allow breeders to relax their feeding efforts and preserve their own energy resources. In Hatchwell's study, males with helpers were 20 percent more likely to survive to the next breeding season than males without.

So many excellent studies have accumulated on helpers that I couldn't possibly review them all here. Charlie Cornwallis, Ashleigh Griffin, Stuart West, and their team identified 154 species of birds with helpers and did a comprehensive analysis of exactly which environmental factors favored helping.[15] Griffin and West, both of Oxford University, are world leaders in using thorough analyses of many species to understand social behavior, including its evolution and its benefits for participants. They have explored kin selection theory in organisms from birds to microbes in a series of brilliant papers.

West managed to get my husband and me an invitation to spend fall of 2016 at the University of Oxford's Magdalen College, where he was affiliated at the time. There we had our own apartment overlooking the botanic gardens on the edge of a rediscovered ancient Jewish cemetery. We put on our black professorial gowns to cross the street to the dining hall, assembling with the other faculty in a side room and then walking in together. As we proceeded through the long and elegant room, the undergraduates stood and waited for us to reach the high table, where the faculty ate. It was all very reminiscent of Hogwarts! Perhaps my favorite memory was sharing what they call dessert in another room, where we passed chocolates, dried fruits, nuts, and even snuff around a semicircle in front of the fire. Actual dessert, which they call pudding, had been devoured earlier at the high table. But it was at this fireside dessert that we devoured long, relaxed conversations each evening.

How to describe Ashleigh Griffin and Stuart West? Griffin is from Scotland while West is from the southern English coast. They both have a marvelous sense of humor, and they simply delight in scientific discovery. Just thinking of them makes me want to go back to Oxford.

Much as Robert Trivers, decades earlier, considered the implications of kin selection, Griffin and West have developed an eye for nature that allows them to discover kin selection patterns where no one else has. The first of their ideas I will discuss here has to do with habitat and cooperation.

Cooperative animals are more likely to live in harsh, hot environments with unpredictable rainfall patterns (this is true, for instance, for the fairywrens in Australia). Early biologists assumed that the harsh environment *caused* cooperation, but it could just as well be true that birds that had already evolved to cooperate were better at thriving in harsh environments. The only way to determine which came first, as it were, was to carefully compare related species in light of what we know about the evolutionary history of birds. We cannot ever know exactly how a long-extinct ancestral species lived, but if all its descendants live a certain way, it is likely that their common ancestor did, too.

Charlie Cornwallis and his team addressed the question of whether cooperative living among birds came before or after they inhabited harsh conditions using 4,707 bird species, all of them cooperative breeders.[16] And they found that cooperative breeding and helpers evolved before the species started to live in harsh environments.

They found that cooperation is most likely to evolve in benign environments under monogamy. From a kin-selection perspective, this makes sense. Under monogamy, a helper will support its full siblings, which are just as genetically valuable as its own progeny. Once helping evolves, these groups are more insulated against harsh conditions than breeding pairs. After all, they have extra sentinels to guard against predators and extra foragers to bring food to the young.

Michael Griesser and his colleagues, including Carlos Botero, my own friend and former Washington University colleague, took a different approach to studying the origins of helpers and cooperative breeding.[17] The researchers viewed the origin of helpers as a two-step process, first requiring extended family life before helpers could arise. The young of some species stuck around past the stage where they were dependent on the parents for food, something called delayed dispersal. These families of parents and their young could remain together for a while as family units. After this critical first step, the scholars postulated that those young that remained with the parents began helping them raise their young. Thus, in their study, they looked at the conditions that favored the evolution of these two steps.

In their data, from 3,005 species, the first step (prolonged family life) was more common in fertile and seasonal environments where remaining together was not particularly costly. These environments had variable rainfall, had variable productivity, and were more open. The family-living birds tended to be larger and more sedentary and to have more specialized diets than the other species in the area. In the second step, helping, the data showed that birds with helpers lived in environments where there was plenty to eat in most but not all years. By breaking down cooperation into two steps, Griesser and his colleagues could show that family living arose in habitats where there was plenty of food, and then helping evolved when the food availability was variable and helping could be important in unpredictable bad years.

As we have seen, helpers can gain two kinds of fitness benefits. First, they can help related breeders raise more young (passing along their genes in the process). Second, they are in groups in a secure environment and can possibly inherit a breeding position if they stick around. If they are helping, they should recognize and favor closer over more distant relatives. Griffin and West also studied this and found a clear pattern: Sure enough, those birds that helped the most were most likely to direct that help to closer relatives.[18] The species in which helping resulted in the greatest number of additional young was also one with a great bias toward helping kin—the Pied Kingfisher. At the other extreme was the Laughing Kookaburra, a species in which nonparental group members neither helped much nor were closely related to the nestlings in the group. It is worth discussing these two species, both of which are actually kingfishers, creating a sort of satisfying set of bookends to Griffin and West's helping-and-kin-recognition continuum.

Pied Kingfishers are common in many areas of the Old World tropics. A striking black-and-white bird, they nest in sandbanks. As I mentioned earlier, I saw them along a tributary of the Brahmaputra River when I visited Kaziranga National Park in Assam, India, with my friends Krushnamegh Kunte and Deepa Agashe. But I only watched and admired their fishing dives. Heinz-Ulrich Reyer studied Pied Kingfishers' helping behavior on Lake Victoria and Lake Naivasha in Kenya.[19] He said the birds were most common along the edges of lakes or rivers, where they can dig their burrows in the soil embankments. Their nests tend to be in colonies, with the burrows 5 to 20 feet (1.5 to 6 meters) apart. They eat fish almost exclusively.

Though there are as many young males as females, among the adults there are nearly twice as many males, perhaps because of greater female mortality during the breeding season. Since the female does all the incubating at night and most of it during the day, she is more likely to fall prey to lizards, cobras, or mongooses, or to be killed if the nest site floods or caves in. Reyer lost few birds during his study, but all of them were females. Another reason for higher female mortality could be that females are ones who disperse, leaving the area, and that brings more risks than remaining in the natal colony.

Pied Kingfishers have two kinds of helpers, primary and secondary, and both are males. The primary helpers are sons of the breeding pair. Secondary helpers are typically unrelated to the adults they help. Primary helpers are allowed to join a pair from the beginning, while secondary helpers are not allowed to join until after the eggs have hatched. Reyer describes seeing young males hoping to become secondary helpers fly up to and try to join one pair after another until a pair finally allows him to stay.

Primary helpers help a lot more than secondary ones do, in part because they join the pair earlier and have a longer time to guard against predators and feed the primary breeders. They also work harder for the family. Primary helpers guard the nest and bring in as many fish (and as many calories) as the mother and father do. The secondary helpers guard less and bring in less than a third as many calories a day as the other caregivers (in part because secondary helpers are more likely to bring in small *Engraulicypris* fish that school closer to the water's surface than the larger, more valuable cichlids do).

Reyer wanted to know how helping impacted a bird's likelihood of surviving to return to the colony the following year. Breeding males with more than one helper were more likely to survive to the next year than those with one or fewer helpers. They were also more likely to survive than primary helpers, but less likely to survive than either a secondary helper or a male not helping or breeding. Females with more than one helper were the most likely to survive to the next year.

According to Reyer's study, nests with no helpers at all produced about two fledglings, while those with primary helpers had twice as many. Those with both primary and secondary helpers had nearly three times as many fledglings as those with no helpers.

Now we can look to the other end of the continuum from the Pied Kingfishers, to the Laughing Kookaburras Sarah Legge studied in Australia.[20] Laughing Kookaburras neither favor kin nor help feed the nestlings, though young birds do join nesting pairs.

I saw Laughing Kookaburras about ten years ago in Mowbray National Park, north of Cairns, Australia, where I was attending a conference. If memory serves, I even heard their legendary call, though they did not sing the song every child knows about the kookaburra in the gum tree. (Apparently, their eerie laughing song is often used as a sound effect in movies set in jungles, though they are not jungle birds.)

Legge studied the behavior of Laughing Kookaburras in a eucalyptus woodland near Canberra. She could recognize individual birds from colored leg bands, unique plumage, and tags affixed to a fold of skin in front of their wings.[21] Andrew Cockburn described Sarah Legge as "seriously brilliant." He told me that he considered continuing her kookaburra study once she had moved on, but did not because he "quickly realized it would take about three or four ordinary mortals to keep the population [study] going."

Laughing Kookaburras are monogamous and do not have young from other parents in their nests, as far as Legge could tell from looking at the DNA markers of 140 nestlings. Their helpers can be female or male, though males are more common, since females disperse when they are two while males do so when they are three. Stepfamilies are also common, mostly forming when female breeders die, so Legge had a lot of variety in relatedness among helpers and breeders to examine.

Unless the first clutch is lost, Laughing Kookaburras

lay only one clutch a year, breeding in the austral spring and summer, September to February. Most clutches have three eggs and are laid in elevated tree hollows. Legge expertly used single rope climbing techniques to get to the nests, which could be more than 30 feet (9 meters) off the ground. The eggs are laid between one and four days apart and incubated for a little over three weeks, beginning with the first-laid egg. This kind of incubation produces a larger chick from the first-laid egg than the young from later-laid eggs. Nearly half the nests lost the youngest nestlings, generally because of aggression by the older siblings. Other young nestlings died later, of starvation, because they could not compete with the larger chicks for food. In all the nests Legge studied, a third of the chicks died before fledging. The young left the nest around three weeks of age and were fed mostly by their parents but also by helpers for nearly two months.

Legge watched the nests and activities of the breeders and helpers long enough to understand their relative roles. Breeders incubated more than helpers did, with male breeders doing more than female breeders, and male helpers doing more than female helpers. Breeders, more often than helpers, brought food to the young, but helpers tended to bring larger, more valuable food items. Male breeders brought the most food, followed by female breeders, who brought in a little more than half the food the males did. Helpers brought less than breeders, with male helpers bringing in over half of what the female breeders brought in, and female helpers bringing in even less. All birds brought less food when there were fewer young or when there were more helpers.

Sometimes helpers brought food to the nest, vocalized,

but then failed to give the food, which was visible in their beaks, to the young. They might never enter the nest, or they might enter and leave it with the food still visible. Young female helpers were most likely to behave this way, during 20 percent of their nest visits. This is a behavior we will see more of in chapter 10, when we meet the White-winged Choughs, an extremely social Australian bird.

Returning to the big picture across all birds, it is clear that helpers bring in enough extra food to make things easier on breeding pairs, but it would be nice to know if this actually means breeders with helpers live longer. Philip Downing, Ashleigh Griffin, and Charlie Cornwallis reviewed studies for such an effect, figuring that adults with helpers could work less and therefore live longer (whether because doing less work preserved their resources or because it meant fewer chances of falling victim to predators).[22]

At first glance, the relationship of helping to longevity is complicated. There is a lot of variation among bird species. For example, Red-cockaded Woodpecker breeders live longer when they have more helpers. But there is no effect of group size on longevity in Karoo Scrub-robins, and Green Woodhoopoe breeders with more helpers have *shorter* lives. Are the differences the result of so-called helpers not actually helping? One way to answer this question is to look at survival of the breeders across more species.

Downing and his team found studies on twenty-three species that included survival data for male and female breeders. Overall, they calculated that helpers increased breeder survival by 8 percent. The pattern was the same for male and female breeders. Breeders generally fed the young 25 percent less when there were helpers. The less the breeders fed, the more likely they were to survive to the next year.

Helpers could also help breeders survive by alerting them to predators.

In general, as we have seen, the more that helpers help, the more closely they are related to the brood. But helpers may have few other places to go, may benefit from hanging around, and may ultimately even inherit the territory and the opportunity to breed. So raising younger siblings or other relatives is not the only way helpers (indirectly) pass on their genes to the next generation. Either way, it is generally true that despite the advantages to helping, it is better to breed oneself and be helped.

In the last two chapters, I will take us into even more complex social circumstances. First, we will look at cases in which more than a single pair of unrelated birds contributes eggs to a single nest; then we will consider the largest, most complex kinds of birds' social groups.

CHAPTER 9

Communal Nesters

Confusion in the Nest

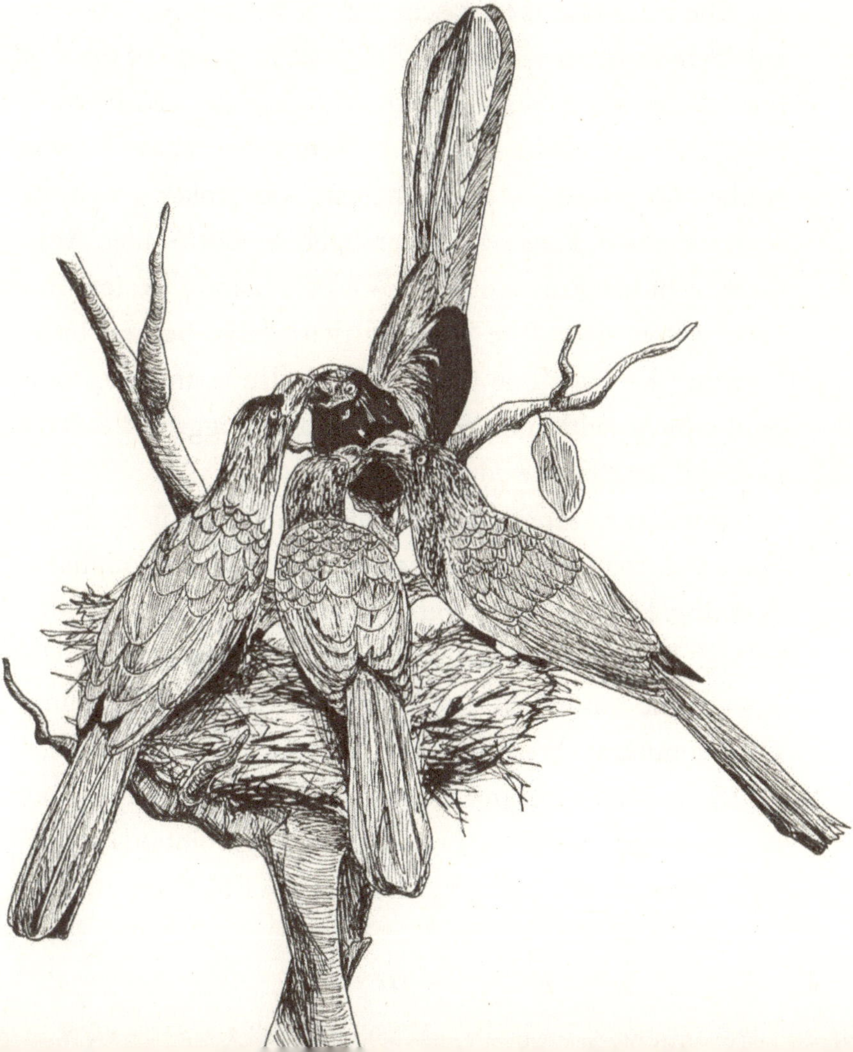

My husband, David, and I were in Gamboa, Panama, hoping to collect a few grains of soil to isolate the social amoebae that we study, but we had not succeeded in getting the proper government permits. Unable to do our collecting, we instead watched birds and enjoyed the company of our friend Allen Herre, who invited us to stay with him in his old wooden Canal Zone home. It was hot, the end of the dry season in March, and we celebrated every breeze. Often, we lingered over lemonade at an air-conditioned hotel. It is wonderful to collaborate with my husband on social amoeba research, and bird-watching for pleasure.

The birds that most fascinated me in Gamboa were the anis, which we saw sitting on branches in groups of three or four. These black birds have dark eyes, large and humped bills, and long tails. My friend Sandy Vehrencamp, who studied Groove-billed Anis in Costa Rica, told me that the birds we saw in Panama were probably Smooth-billed Anis; Groove-billed Anis are rare in Gamboa, and Greater Anis have yellow eyes. The story of these large-beaked birds' breeding habits is one worth telling. These are true communal birds, with multiple females laying eggs in the same nest and tending the young together.

It will not surprise readers who've made it this far in the book that there are many open questions about communal breeding. How do the parents divide the work of tending the babies? Wouldn't natural selection favor cheaters who laid eggs in the communal nest but did not care for the young? Or is communal breeding a short-lived evolutionary fluke?

Soon after I became interested in the anis, I ruled out the idea that communal breeding was an evolutionary novelty. After all, the crotophagine cuckoos (the three ani species and the Guira Cuckoo) that share this communal breeding system diverged from the other New World cuckoos, including their closest relative, the roadrunner, several million years ago, and they are still going strong.[1]

Some might think that birds that breed communally are not so different from those with helpers (discussed in chapter 8), but they have some important distinctions. Philip Downing and his team showed that helpers tend to arise in lineages with monogamous mating, while communal breeding arises in lineages with polyandry, where females mate with multiple males and the males do the parenting.[2] An important difference between communal breeders and helpers is that helpers tend to be related to those they help; communal breeders are typically unrelated to one another and so must get a direct benefit from communal breeding. But it is complicated, since communal breeders can also have helpers that are related to one of the breeders. To explore communal breeding, I'll start with the Groove-billed Anis.

Sandy Vehrencamp is the best kind of biologist, one who delves deeply into the organism she studies while also having the insight to derive general theories from her study. She is perhaps most famous for developing what we call "skew theory." This theory predicts that when individuals reproduce in a group, they do not necessarily do so equally. One individual may lay most of the eggs or sire most, if not all, of the young. Even so, being in the group can be the best option for all. Where an individual falls in a hierarchy depends on things like an individual bird's condition relative to that of the other group members, and whether the bird finds it advantageous to be in a group at all. Besides skew theory, Vehrencamp is well known for her book *Principles of*

Animal Communication, which she coauthored with her husband, Jack Bradbury, and which won the Exemplar Award, the Animal Behavior Society's top honor for books.[3]

Sandy Vehrencamp and I have a lot in common. We both do at least some of our research with our biologist husbands. And we both began our careers with solo-authored papers in the prestigious journal *Science*. That can be a career-defining event (I have no doubt it got me my professorship at Rice University, and I imagine Sandy's single-author *Science* paper was similarly beneficial to her career).

Vehrencamp studied Groove-billed Anis in Guanacaste Province in Costa Rica.[4] She told me she had a hard time catching and banding the birds until she hit upon the idea of using a tame Groove-billed Ani that she could temporarily place in a walk-in trap so that others would follow. She could then band and release the birds.

She was curious about communal breeding because many researchers were surprised that there was *any* cooperation among nonrelatives. When she was working in the late 1970s, kin selection was increasingly appreciated as a motivator of cooperation, following Bill Hamilton's original work on inclusive fitness[5] and Richard Dawkins's landmark exploration of the topic, *The Selfish Gene*.[6] Vehrencamp realized that young dispersed in early adulthood, so the individuals forming communal groups were likely to be unrelated. Nonetheless they evolved to cooperate in a communal nest because each gained more than they could have gained nesting alone—though the dominant pair gained the most. Remember, natural selection does not dictate that cooperation has to be *fair*, only that it is better than the alternative for a given individual. Vehrencamp wanted to understand what exactly those advantages might be.

The nesting groups of Groove-billed Anis that Vehrencamp studied had between one and four pairs laying eggs in a communal nest—a large, open-cup nest made of twigs and placed either high in a tree or low in marshy plants. As far as she could tell, the pairs remained monogamous, so these were communal groups of pairs rather than mating free-for-alls. Each group was ruled by one dominant female whose mate did much of the egg incubating (including nearly all of it during the most dangerous nighttime period). He also fed the nestlings about twice as much as the other birds did. The dominant female did little incubating herself. Instead, the lowest-ranked female was the main daytime incubator. (Interestingly, Vehrencamp sometimes found only one pair tending the nest, so Groove-billed Anis do not always nest communally.)

Though the birds in each group cooperated, not all was calm—eggs were frequently ejected from the nest. Vehrencamp wanted to know which eggs were tossed and who ejected them, so she watched a three-female nest continuously for twenty days, then extended the study with marked eggs to another fifty-four nests. She found that a female pushed eggs out of the nest only before she herself had laid any eggs. This suggests that while the female cannot tell which eggs are hers, she knows that when she has not laid eggs, they cannot be hers. This means that the first female to begin laying loses the most eggs, and the last female to begin laying loses no eggs because egg tossing stops when everyone has laid eggs. This *last* female, then, is the dominant female.

But there is another element to the story: The last-laid eggs hatch later because the adults start incubating the eggs after the first two or three are laid. Since chicks are fed the moment they hatch, the young that hatch later will be smaller than their older nestmates. More than half of the larger young survived, while less than a quarter of the smallest nestlings did so. Thus it is clear that late laying, like early laying, has a cost, and the last to lay should not delay for fear that her young will be the smallest. Vehrencamp surmised that more dominant birds may remove early eggs to make it harder for other birds to have the resources to keep laying eggs. They simply won't have the fat and protein reserves to make more eggs. Overall, Vehrencamp noted that all females succeeded in having some young raised in the nest, though the dominant female had the most.

In Groove-billed Anis, if the first nest is destroyed by a predator and the breeders have to start over, the same scenario generally plays out, with the same birds starting early and the same ones laying later and more often. Those successful later layers tend to be older and in better condition. This pattern does not hold in other closely related species. In both Greater Anis and Guira Cuckoos, a different female is likely to become the first layer in a renesting attempt.

It is common among graduate students searching for a thesis project to feel like all the questions have been answered and to despair of ever finding a new topic to study. I certainly felt that way myself, but that is a story for another time. I know I would have felt that way had I wanted to study Groove-billed (or really any) Anis after Vehrencamp's excellent work, but I would have been wrong. There is always more to learn.

Christie Riehl shows us as much with her fascinating studies of Greater Anis, which are even more social than Groove-billed Anis, because pairs never nest alone. Riehl, now a professor at Princeton University, chose Greater Anis to learn more about communal breeding. I was happy to hear from Christie how much Sandy helped her get started, sharing ideas and answering questions, just the way scientists should.

Unlike Groove-billed Anis, Greater Anis never nest in pairs; they are always communal. Greater Anis are abundant in the right habitat, but they are difficult to study because their nests are built along shorelines, either in branches overhanging water or in vegetation actually growing in the water.[7] They have to be studied from a boat. Perhaps because of this, the breeding behavior of Greater Anis was little known before Riehl's landmark work along the shores of the Barro Colorado Nature Monument in Panama.

Riehl and her team banded Greater Ani adults by capturing them with mist nets at communal roost sites during the nonbreeding season or near their nests in the breeding season. They mounted the nets on aluminum poles in the shallow water parallel to the shore and had to extract the birds from the nets—hard enough to do on land!—while precariously standing in kayaks. From their boats, they watched the nests and the birds' displays, marked the eggs, and set up motion-detection cameras to record behavior and predators.

Riehl and her team first wanted to know how many birds were in a nesting group. They could tell when the females were about to lay eggs because at that point the birds lined the nests with green leaves.

Riehl and her team found that 75 percent of the eighty-seven groups had two pairs, 20 percent had three pairs, and only 5 percent had four pairs. Some groups also had helpers, typically male young from previous years. The members of a group stayed together, sometimes even over ten years. The longer they were together, the more efficient they were at managing nesting synchrony and the more success they had.

When the eggs were laid, they had a white chalky coating, called vaterite, that rubbed off during incubation to reveal a blue shell underneath. The eggs were huge, like all cuckoo eggs. Females laid between three and seven eggs, averaging about four.

Some of those eggs would be tossed out of the nest by other females. The first-laid egg was almost always ejected, and in general, early-laid eggs were more likely to be tossed than later-laid eggs. The more pairs in the group, the more ejected eggs. Four-pair nests had the most eggs ejected, and the adults ultimately abandoned the nest and left the group. In nests with two or three pairs, once there were two or more eggs in the nest for a couple of days, egg tossing ceased.

Nests mostly failed because of snake predation, and Riehl surmised they might have been safer farther out over the water. But in her experience, the larger the group, the less chance of predation.

Among anis, rogue females also fly in and lay eggs without being part of the group at all. The females that do this generally have just lost their own nests, so they fly to a neighboring communal nest and pop in an egg.[8] This parasitism happens in roughly 15 to 30 percent of nests. It would seem like a big problem, but it turns out these eggs are often laid at times that will not result in babies, either because they are laid too early and jettisoned or they are laid too late and do not hatch during the incubation period—after which the imposter's eggs will get cold and die. These birds and other bird species in the cuckoo family have some of the fastest development times of all birds, so a bird trying to lay an egg in a stranger's nest will probably lay too late.[9]

The two other crotophagine cuckoos, the Smooth-billed Anis and the Guira Cuckoos, also nest communally. Jim Quinn has worked on both. Quinn is unmistakable, with a snowy white beard and a halo of hair. He and his team study Smooth-billed Anis at the Cabo Rojo National Wildlife Refuge in southwestern Puerto Rico. I have not been there, but when I was eight, my family and I lived in a two-room apartment in San Juan, Puerto Rico, for a few months while my father studied the economics of manufacturing. We visited Ponce in the south and El Yunque rainforest in the east, but not the dry forests of Cabo Rojo. We had previously lived in Mexico, where, at school, I had learned Spanish well enough to understand the Puerto Rican accent that stymied my parents. I even translated for my mother when we went grocery shopping.

In order to capture Smooth-billed Anis for banding, Quinn and his team either used a lure bird in a trap, as Vehrencamp did, or caught them in very high mist nets. Smooth-billed Anis also nest communally. The females also toss or bury eggs laid before they themselves start laying.[10] They have to lay a lot of eggs to make up for the ones they lose, averaging slightly over six eggs per female. Each group averages about six individuals, and the birds seldom attempt to renest in a season. Like the Groove-billed Anis, and unlike the Greater Anis, the male that takes on the risky job of nocturnal incubation is mated to the female that lays the most successful eggs.

Smooth-billed Anis have alarm calls that distinguish terrestrial from aerial predators.[11] They make a sound like *chlurp* in response to raptors, which causes the anis to dive into the bushes. If they detect a ground predator, they make a sound like *ahnee*, and the birds fly to higher perches. Apparently Greater Anis have a similar system, and this is another benefit to being in groups.

The remaining species in the subfamily Crotophaginae is not black like the anis but colored more like its distant relative the roadrunner. Guira Cuckoos are tan underneath with brown backs delineated with white lines. Most notable are their ginger head feathers, which stand straight up and give the birds a surprised look. Yellow-orange eyes and similarly colored beaks complete their appearance.

I saw Guira Cuckoos in Pirenópolis, Brazil, during an Animal Behavior Society meeting. The ones I saw were always in groups, sitting high on branches as close together as could be. I was excited to connect them to all the work my Brazilian friend Regina Macedo did in the early 1990s as part of her Ph.D. dissertation at the University of Oklahoma. Now retired from biology, she is a well-known artist who paints colorful illustrations of people and natural landscapes.

At the time Macedo studied them, Guira Cuckoos were common near her home in Brasília, but that is no longer the case, as the suburbs have spread. Guira Cuckoos nested in groups that averaged about six birds, or three pairs, but these pairs were not particularly monogamous; half siblings were more common than full siblings in the nests. Though there was little commitment to one mate, mating tended to be restricted to within the group. Fathers were more related to one another than mothers tending the same nest, reflecting the lower male tendency to disperse.[12]

Like the female anis, Guira Cuckoo females eject eggs, particularly early in the egg-laying period. Even after hatching, young (usually the youngest chicks) are sometimes killed, whether tossed directly out of the nest or carried some distance from the nest and then dropped. This behavior is hard to understand, because it is not clear, at least to us, how the perpetrators recognize that these are not their own babies. It could be that females that laid few eggs are doing it to speed up a new nesting attempt where they might lay a larger number of eggs. Or possibly there are more eggs or young in the nest than can be successfully reared. Overall, Guira Cuckoo family life is complex.

The Crotophaginae are among the best-studied but not the only communal breeders. Though communal breeding is not common in perching songbirds, it is found in the Taiwan Yuhina, studied by Sheng-Feng Shen and his team.[13] I saw this bird after a week spent at a forest station in Taiwan giving lectures on social behavior and evolution. The colleagues that invited me and my husband, David, along with Stuart West, to the forest station kindly hired a bird guide for us, and we went up to the Dasyueshan Forest Road.

Taiwan Yuhinas are not rare where we saw them—in fact, they are the most common bird in the mountains of Taiwan. They chattered in the trees, moving here and there in small flocks. The birds' crest, chocolate edged with black and standing nearly straight up, was much more exaggerated than that of the Tufted Titmouse I see at home. These birds are so adorable and seem almost tame, so it is no wonder Sheng-Feng Shen decided to study them. His team observed them at National Taiwan University Highland Experimental Farm at Meifeng, 7,055 feet (2,150 meters) elevation. In this high-altitude subtropical cloud forest, many days were foggy and rainy, which made it difficult for the birds to forage. The vegetation was mostly orchards and ornamental trees surrounded by natural beech, oak, and laurel forests. The birds built their nests 8 yards (7 meters) aboveground, on average, and out on the thinnest branches, so researchers had to use a cherry-picker truck to reach them.

Because the Taiwan Yuhinas are altitudinal migrants, breeding at over 7,000 feet (2,130 meters), where the climate is often cold, wet, and variable, researchers can compare these birds' cooperative actions in poor, excessively rainy and cold seasons and in good, warmer breeding seasons. Cooperation may be more necessary in poor weather,[14] and groups allow for speedier rebuilding after nest failure.

Groups in Sheng-Feng Shen's study typically had four birds—two pairs—sharing the same nest, though there could be up to seven birds in a group; rarely, there were only two.[15] About a third of the groups contained an unpaired bird, usually a male. Most of those unmated helpers were progeny of a group member from an earlier breeding season. Shen and his team used DNA markers to verify that nearly three-quarters of the young were the progeny of social pairs. When the female's young were not sired by her partner, they were usually sired by a male in the same group. Subordinate females were more likely to mate outside the pair bond. The pairs that made up a group of breeding Taiwan Yuhinas were generally unrelated to each other.

The birds in the groups Shen observed stayed together for multiple years. There was a dominance hierarchy within the sexes, with those lower on it deferring to those higher. This means they do not have to fight each time they approach a resource because they know who is on top. It is an efficient system and common among animals. There were a lot of within-sex chases and dominance signals. For example, the third-ranked males did not sing when the higher ranked ones were present unless they were having an altercation with another group about a territory boundary. Generally, the top-ranked male sang more than the others.

Nest failure rates in Taiwan Yuhinas are high, around 77 percent, due to abandonment, predation, and severe weather, according to Shen and his team. Feeding the babies is typically done in groups, with more than half the group arriving at the nest at the same time. This feeding synchronization is particularly intense at more exposed nests. If these nests were to be visited by adults more often, predators would be more likely to learn their locations. Another consequence of synchronous feeding was that the less competitive young could also get food.

Female Taiwan Yuhinas compete in several ways within the cooperative framework of breeding together and caring for young in the same nest. For example, females try to block other females' access to the nest to lay eggs, a behavior Shen calls "tussling." Females also try to lay more eggs than the others, and they begin to incubate day and night when they have laid their own eggs, attempting to ensure theirs will hatch before later-laid eggs and will have better access to food.

Females tussle early in the morning, just after the males begin singing. Tussling happened in all but one of the thirty-six nests Shen and his team watched. Of the sixty-one tussles they saw, one female stopped the other from laying less than half the time. Tussling happened most aggressively before incubation had started, when the young that hatched would be competing for food.

During rainy periods, fewer eggs were produced, but more of them survived.[16] This meant that the total number of surviving offspring from the group was lower when conditions were good and higher when it was cold and rainy.

Shen and his team attributed this mostly to the fact that all the eggs were laid at the same time by the different females in cold periods—the result of less fighting. This story, uncovered through careful research, illustrates the cost of competing for reproduction.

Another interesting and well-studied communal nester is found in New Zealand. Pukekos are a subspecies of the Australasian Swamphen. The name swamphen makes a lot of sense since they look a bit like their distant chicken relatives and walk about on long-toed, chicken-like feet. Pukekos are in the Rallidae (or rail) family, which is close to the family chickens are in (Phasianidae). Like other rails, they do not have webbed feet, though they are largely aquatic. Pukekos have pink to red legs and a fleshy red shield that goes up between the eyes. They are brown to black with a lovely navy breast and neck, as if they're wearing navy suits with black jackets. They have pure white feathers under their tails.

I like to connect birds in my mind so that I can keep them straight. The three birds I think of when I try to mentally place the Pukeko are the coots, Common Gallinules (formerly called Common Moorhens), and Purple Gallinules. I used to see these birds regularly near Houston, Texas, at Brazos Bend State Park, though the Purple Gallinules were something of a treat. I'd see Eurasian Coots screeching across the water early in the mating season as they fought for territory on Herthasee in Berlin.

But just because these birds *look* like Pukekos does not mean they are necessarily closely related.[17] Pukekos live in groups of several males and one or two females. Like other communal birds, the female Pukekos lay eggs in a common nest and jointly care for the young along with the males. But it is much more of a free-for-all than with the Groove-billed and Greater Anis, where each female is paired with one male. Instead, breeding Pukekos in a group all mate with one another. And I mean *all.* Females mount females occasionally and males mount males. It is even possible for females to exchange sperm if there are cloacal contacts between females and they have recently mated with males, suggests John Craig, one of the early researchers of Pukekos. It might sound like a love fest, like the peaceable bonobo chimpanzees are known for, but that is not quite true, since there is a strict dominance hierarchy and competition for mating. Subordinate males that mount dominant females are likely to get pecked by the dominant male for doing so.[18] Still, all or nearly all the males in a group mate with at least one female, and all help incubate and rear the chicks, something that can stabilize cooperation since everyone has a stake in the nest.[19]

Just like the ani chicks, Pukeko chicks do not hatch all at the same time. Incubation begins before the last eggs are laid, so the first chicks to hatch get fed and grow before the later chicks emerge. And those early chicks tend to come from eggs laid by the dominant female. This advantage extends throughout the early chicks' lives, as these chicks are more likely to become dominant in their own nests.[20] Later-laid eggs are in fact less likely to hatch at all.[21] Laying asynchrony is more common when there is more than one laying female.

One might even wonder why secondary females are tolerated, but it turns out that the males spend more time incubating when there are more eggs, and they might abandon the nest altogether if there are fewer eggs.[22] It might seem odd that a male would ever abandon even a small nest. Indeed, they would only do so if there were better alternatives. Besides smaller nests bringing fewer progeny into the world, there is another challenge to small groups of Pukekos in the area. These birds are highly territorial, and larger groups are better able to defend their territories.

Territories are crucial to the success of the Pukekos because the young leave their nests very early, a couple days after hatching, though they are still entirely dependent on the adults of the group, who guide them in smaller groups to temporary nest-like structures. Having multiple smaller groups reduces the likelihood that a predator could find and gobble up all the young in different groups at once.

Like many social organisms, Pukekos have dominance badges, something that identifies an individual's rank visually. These badges signal rank and make resolving disputes among individuals in a social group more efficient.[23] In Pukekos, that badge is the frontal shield on the face, and its size signals rank. The fleshy structure extends from the bill to up between the eyes, and it is bright red. Cody Dey and his colleagues measured the shields and then decided to manipulate their width to see whether smaller shields resulted in lower rank. They did this during the breeding season at Tāwharanui Open Sanctuary in New Zealand. They did not hurt the birds, but simply put a little bit of black paint on the edges of the shield so that it would match the feathers and appear smaller. (Controls had their shield edges painted with red paint that matched the shield's natural color to make sure it was not just the handling alone that caused results.) Dey and his team observed the birds, then recaptured the males a week later and remeasured their actual shields. By then the paint had mostly worn off.

As they expected, birds with larger shields were more

dominant. The birds that had their shield size reduced received more aggressive attacks, challenges, and wings-up displays from others in their group. Astonishingly, when at the end of the study the team measured the shields that had been painted smaller, they were actually narrower than they were at the start. Apparently, the birds responded physiologically to status signals from other birds. The dominance hierarchies Dey and his team studied help Pukeko groups get on with the tricky business of communal reproduction rather than fight all the time; once the hierarchy is established, subordinates do not challenge dominants for things like access to resources.

The anis, Guira Cuckoos, Taiwan Yuhinas, and Pukekos are all common birds, so clearly communal breeding can be successful. Yet it has not evolved often in the ten thousand species of birds, perhaps because of the intricacies of working out cooperation among nonrelatives who must, necessarily, each get their share. It is hard for individuals to manage their own genetic interests when all lay their eggs in a common nest that cannot hold an infinite number of eggs.

CHAPTER 10

Supersocial Groups

Birds That Are Always Together

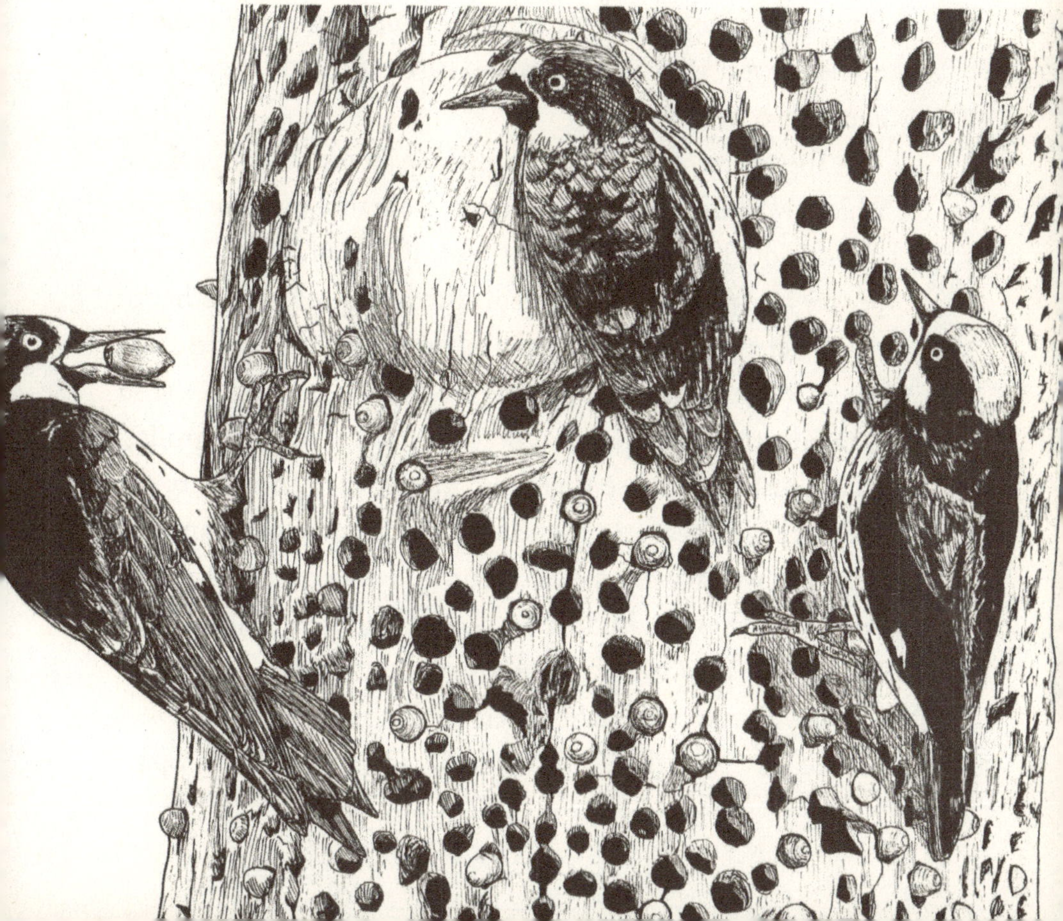

Out of the canyon they came screaming, improbable blue catching the sun against the red Wingate Sandstone. There was a flock of about twenty, then another, and finally a larger group of thirty. They followed the same path, staying close to the cliffs, calling to one another all the way. A lone bird flew behind the groups, calling constantly, seemingly desperate to catch up.

They were Pinyon Jays, I realized with a shiver, the magical bird that my friend Kris Johnson has devoted her life to. The birds my daughter and I saw were a winter flock, flying out of the canyon toward the bird feeders in Fruita, Colorado.

Pinyon Jays are exquisitely social in nearly everything they do, making them worthy of being called *supersocial.* This loose category covers birds that have multiple ways of being social, making their lifelong group living stand out. Pinyon Jays, for instance, nest in colonies, forage together, cache pinyon seeds together, and care for their fledglings in a joint crèche.[1]

The supersocial birds also include Acorn Woodpeckers, which, working together in groups, store thousands of acorns in individual holes in granary trees. Bands of Acorn Woodpecker brothers and societies of sisters also come together to mate, defend their granaries, and raise their young. In Australia, White-winged Choughs are never alone. They are so dependent on groups that they will kidnap other groups' young just to increase their own group size. Then there are the Sociable Weavers of southwest Africa, whose communal nests can weigh a ton, yet each family has its own nest chamber that it shares with helpers. Together, these four birds illustrate the breadth of complex sociality in birds.

If you saw a Sociable Weaver foraging, you might not realize it's a remarkable bird. It looks a lot like a House Sparrow, with a large beak, a black chin, and a brown cap. But overall, it is lighter colored than a House Sparrow, with dark feathers edged in tan, giving it a scalloped appearance along the shoulders and sides. The Sociable Weavers are never quiet; they constantly chatter. The chatter sounds all the same to me when I listen to it on recordings from the Cornell Lab of Ornithology's Macaulay Library, but I'm sure the birds are saying more than just "I am here." Gordon Lindsay Maclean broke the calls into eight different sounds: contact, chatter, entry, threat, greeting, begging, flight, takeoff, alarm, and nest calls.[2]

Sociable Weavers build massive nests that look kind of like haystacks stuck up in trees, mostly camelthorn trees.[3] The colonies average seventy-five birds but can have more than five hundred, all of whom work together to build the thatch that overlays the nests, which have entrances on the bottom of the thatch. The roof of the nest is made up of an underlayer of grasses roofed over with thorny twigs.[4] Pairs of birds build separate nest chambers into this substructure of grasses, each with an entrance tunnel opening to a small chamber where they raise young and huddle on cold winter nights. The chambers are lined with grass seed heads, furry leaves, and even pieces of human-made fabric.

The nest haystack is kind of like a condo: Each Sociable Weaver family has its own separate dwelling but all are responsible for the enormous building. Maclean describes the effort put into construction: "All members of a colony build continually, even though some may build more than others. Young birds show signs of a building drive at the age of 80 days. This drive increases until it is so strong in adult Sociable Weavers that they build at all times except when resting, feeding or attending to eggs or chicks."[5] Actually, building is more a male than a female activity, but clearly the nest mass is essential to the success of Sociable Weavers.[6]

The chambers are used all year long, in the austral summer for breeding and in the winter for staying warm when the breeders and all their helpers pile into the same chamber.

René van Dijk and his team, working at the Benfontein Nature Reserve in South Africa, studied temperatures in twenty Sociable Weaver nests from September through December (austral spring into summer).[7] Keeping the birds out so they could not influence the temperature, the researchers put a thermistor probe into empty nest chambers at three different locations in the colony: at the edge, in the center, and in between. They measured the temperature outdoors, near the thatch in the shade but not right on the nest.

Outdoor ambient temperatures ranged from as low as 35 degrees Fahrenheit (1.4 degrees Celsius) at night to as high as 108 degrees Fahrenheit (42.4 degrees Celsius) during the day. Inside the nest chambers, however, temperatures stayed several degrees away from these extremes. This indicated that to some extent the nest structure buffered the nest chambers from ambient temperature fluctuations. The smallest thatches were less buffered, but beyond that there were few differences by nest mass.

Gavin Leighton looked at winter thermal buffering as part of his Ph.D. research at the University of Miami.[8] We met during that time, and I've since followed his career with interest. I even invited him to share his insights on Sociable Weavers at a conference my husband and I organized on what it meant to be an organism.

In winter, Leighton took multiple measures of the temperature in chambers where the birds were kept out, finding that thermal buffering was higher in chambers in the middle of the nest. Proximity to neighboring chambers increased thermal buffering, too. He also found that while larger nest masses buffered better, the benefit tapered off once the nest mass reached about twenty chambers.

Remember, though, these tests showed the buffering benefits of nest structures without the birds. The Sociable Weavers themselves, it turned out, were even better at tempering the winter chill—not least because they are, well, sociable. Matthieu Paquet and his collaborators proved that birds get a great deal of warmth from one another.[9] The average temperature in a roosting chamber increased from about 54 degrees Fahrenheit (12 degrees Celsius) with one bird to about 91 degrees Fahrenheit (33 degrees Celsius) with five birds. At an ambient temperature below 73 degrees Fahrenheit (23 degrees Celsius), the birds had to expend energy to stay warm. But nest cavities with four or more birds could stay above that temperature even when external temperatures were as low as 43 degrees Fahrenheit (6 degrees Celsius). That the birds let only their own family and helpers into the cavities means having more helpers is useful in maintaining temperatures, too.

Another factor that can increase group size is scarcity of resources. When there aren't enough trees for Sociable Weavers to nest in, some colonies can become larger than is optimal. This "clumping" can, in turn, impact food resources, which are likely to become scarcer.

Claire Spottiswoode has studied the trade-offs involved in Sociable Weavers living in smaller and larger colonies. I have met Claire a number of times, and I admire her detailed research and her passion for working outside. She now focuses on African Cuckoos from her research homes at Cambridge University in the UK and the University of

Cape Town in her native South Africa. To consider colony size in Sociable Weavers, Spottiswoode selected a study site and then identified fifteen colonies varying in size from four to seventy birds.[10]

Among her findings, Spottiswoode noted a number of disadvantages in the larger colonies. The birds themselves were smaller and laid smaller eggs, for instance, and the nestlings were more infested with bloodsucking mites (though they had fewer infections, as indicated by an immune challenge test). Adult Sociable Weavers in large colonies lived longer than those in smaller colonies, but their young were more likely to die, either by being thrown out of the nest early or starving after fledging. Larger colonies were also more likely to lose all their young to snakes, mostly Cape cobras and the delightfully named but lavishly venomous boomslangs, which methodically ate their way through the colonies, nest by nest. In all, adult Sociable Weavers lived longer in large colonies, but the young birds lived longer in small ones.

A big question about Sociable Weavers is how they manage to cooperate in building such an expansive structure as their communal nest. What keeps some birds from being free riders and letting the other birds do all the work? Gavin Leighton and his coworkers explored this in the nonbreeding season, watching 127 birds at two colonies for nearly three hundred hours.[11] It turned out that the males spent more time than the females did building the thatch. The scholars used data on relatedness among colony members to discover that males with more relatives at the nest spent more time building and stuck more grasses and sticks into the nest than males with fewer relatives did. Leighton concluded that the birds that build the communal nest do it, at least in part, to help relatives.

That leads us to another question: How do these males end up with relatives in the colony? To understand the mating and parental care system of Sociable Weavers, we can turn to Rita Covas, who works on a population that has been banded since 1993.[12] This group, in South Africa's Benfontein Nature Reserve, is part of a long-term study run by Covas with Claire Doutrelant.

In this population, Sociable Weavers built colonies of ten to two hundred birds. Individual nests usually had two to four eggs in a clutch, and the birds could have multiple clutches in a season (breeding begins in September, with the rains). The parents continued to breed as long as they could find enough insects to feed the young; in practice, that meant as long as the rains continued.

Breeding in Sociable Weavers is a story of monogamy. Pietro D'Amelio and his team used data from ten to fourteen colonies per year between 2008 and 2019.[13] The team caught all the adults in mist nets placed around each colony or in a hand net at a specific nest entrance. They banded the birds and took blood samples to determine sex and parentage. Then they analyzed data from 703 pairs, finding that males and females stayed together. Only 6 percent were found with a new partner when the previous partner was still alive. Additionally, out of 1,275 clutches that had more than one egg, only 5 percent had eggs from more than one male (this amounted to only 25 of the total offspring).

The Covas team learned that, in addition to the parents, breeding Sociable Weavers often had helpers. These were usually male progeny from earlier in the season or from previous years, and they typically numbered one to five per nest chamber. To determine how much helpers actually helped, the researchers collected data on 1,598 adults and nestlings at nineteen colonies.[14] They put a plastic number tag at each nest entrance so that they could keep track of who fed where. In the first year, 30 percent of the breeding pairs had helpers, while in the second year, 82 percent of them had helpers. Any given pair could attempt to breed as many as eight times a year, though five times was the average. Overall, groups with more helpers successfully fledged more young, but larger colonies fledged fewer young. Further, the parents were likely to be in better condition when they had helpers, because they were less responsible for feeding the young.

Covas found that of the 710 nesting attempts they followed over two years, 77 percent failed, nearly all of them because of predation on the young by snakes. The helpers had no influence here; there was simply nothing the birds could do once a snake reached their colony.

In a more recent study, a team led by D'Amelio found that while helpers generally increased reproduction, their presence could not overcome the ill effects of extreme heat and drought.[15] At 77 degrees Fahrenheit (25 degrees Celsius), the probability that a pair would fledge young was only 25 percent, while for a group of one pair and four helpers, the probability increased to 55 percent. But at 93 degrees Fahrenheit (34 degrees Celsius), the difference was much less: 11 percent for a pair and 15 percent for a pair and four helpers.

To explore helping and the needs of the young in more detail, the Covas group did two kinds of experiments to alter the need for helpers. In the first, they made food more available by scattering millet and red rapeseeds under some colonies while leaving others unsupplemented.[16] They found that the adults and yearlings weighed more in the food-supplemented colonies. They also found that some of the one-year-old birds attempted to breed, something they normally do not do at that age. Covas and team determined that when it is easier to rear young, more birds choose to breed than to help. This meant that average group size at an individual nest chamber was smaller at the supplemented colonies.

In the second experiment, the Covas team added two chicks that were one to three days old to nests with and without helpers or removed two chicks.[17] They never increased brood size above the natural maximum seen in the field—six young—and to ensure the manipulation itself did not have any impact, for controls, they switched two chicks between nests without changing the total number of chicks in those nests. So, there were three kinds of broods: those with two additional nestlings, those unchanged in size, and those with two fewer nestlings. For each treatment there were nests with and without helpers. They watched each nest for four days, and for their analyses, they did not count nests whose contents were eaten by snakes.

Overall, the number of feeding visits per hour was a little higher at nests that had two additional chicks, when compared with control nests that had no change in chick number. The number of visits was significantly lower at nests that had two chicks removed. Fewer chicks survived at the enlarged nests, as compared to the control nests. Helpers were largely responsible for the increased feeding rates at these nests. Overall, more young fledged from nests with helpers regardless of the experimental treatment.

Sociable Weaver nests are prominent—and consequential—in the arid Kalahari landscape. The birds change the surroundings to such an extent that they could be considered ecosystem engineers, like beavers.[18] For example, when compared with the grasslands and the soil under uninhabited trees, the soil under trees populated by the weavers has more nitrogen, phosphorus, and potassium.[19] The birds' presence also increases species diversity. Anthony Lowney and Robert Thomson monitored an area called Tswalu Kalahari that contained 250 Sociable Weaver colonies.[20] Looking only at trees with nest masses of more than twenty nesting chambers, they discovered that, compared with trees without nests, trees with nests contained many times more species. In fact, the larger the colony, the greater the species variety. Other birds roosted in the nests, too, including Acacia Pied Barbets, Ashy Tits, and Scaly-feathered Finches, but that wasn't all. In their paper, they show a gemsbok grazing under a Sociable Weaver colony, a cheetah climbing down from it, and a genet (a small wild cat) foraging on arthropods atop the colony.

Clearly Sociable Weaver colonies are important to many Kalahari Desert animals. Some, like the African Pygmy Falcon, *only* nest alongside Sociable Weavers.[21] These falcons deter snake predation at colonies, which increases nestling survival, but their net effect is not clear, since the falcons themselves snack on nestlings. Kalahari tree skinks are three times more common in Sociable Weaver nests than they are in trees without them,[22] and these lizards are found at nests even when they also contain nesting African Pygmy Falcons, which prey on the skinks. Tasmin Rymer and her colleagues surmise that the skinks prefer the nests because they offer shelter and thermal protection, as well as easy foraging for the arthropods that live within them. One might say Sociable Weavers are supersocial at another level, with their social lives impacting their habitat, too.

Acorn Woodpeckers build their own memorable structures in a very different kind of habitat. These birds can be found in pine-oak woodlands along the western coast of the

United States, along the Arizona–New Mexico border, down both mountain ranges through Mexico and Central America and into Colombia.

When I see an Acorn Woodpecker, I want to laugh, as it looks like a clown, with a red crown and a shiny black band across the top of its face that encompasses the eyes and then runs down its back. It has white above the beak and on the chin, with the beak encircled in black, and a black upper chest with a white belly, rump, and wing patches.

Acorn Woodpeckers are famous for their granary trees.[23] These are usually large trees, typically oaks, into which the Acorn Woodpeckers have drilled as many as fifty thousand holes, each large enough to hold just a single acorn. These granary trees are quite spectacular. The large trunks are studded with acorns protruding from their holes, giving the tree the look of a beaded dress. Besides trees, Acorn Woodpeckers drill storage holes in telephone poles, fence posts, and even wooden buildings. If they run out of holes and cannot drill new ones fast enough, they will dump acorns en masse into places like old nest cavities in hollow trees or even abandoned buildings.

Acorn Woodpeckers are much loved by Californians, including my husband, David, who told me that they were one of his two favorite birds when he was growing up in Southern California. (His other favorite bird was the California Scrub-jay.) He loved Acorn Woodpeckers because of their comical beauty and because they seemed exotic. He saw them regularly at only one place: on a telephone-pole granary on his walk to elementary school, right at the intersection of a railroad track and a busy street in South Pasadena. Walt Koenig spent decades studying them at a more natural location, Berkeley's Hastings Natural History Reservation.

Walt and I are near contemporaries and so have seen each other at animal behavior and behavioral ecology conferences for decades. We came at cooperation from different corners of the animal kingdom, me with wasps and amoebae and Walt with Acorn Woodpeckers and other birds, but that did not stop us from becoming friends. Walt has a bemused air about him and the sharpest sense of humor. If a talk has been going on too long, or there is an overly heated academic argument, I can count on Walt for a humorous and kind interpretation. Besides his work on the woodpeckers, Walt has brought the field together with two edited volumes on cooperation focused on birds, books I have relished. He sends out an annual acorn report that summarizes his team's efforts at counting the acorns along the west coast of the United States, along with family stories and diner recommendations.

Koenig and his collaborators have found that Acorn Woodpeckers generally drill holes in their granary trees in the winter, though they will also drill them in the fall if they have gathered more acorns than they have storage holes. The cavities are not easy to drill, so the woodpeckers do not drill them all at once; they begin them and then return to finish later. Once a hole exists, it can serve the woodpeckers for generations, as long as the tree remains. Each year, the woodpeckers insert new acorns into their granary holes. Any given granary tree is in the territory of one group of woodpeckers who guard it from anyone else. Koenig and team estimated that an average group of five woodpeckers might drill about five hundred holes a year.[24] At this rate, it would take more than eight years to produce a modestly sized granary of four thousand holes, and more than a hundred years to make a granary of fifty thousand holes, close to the maximum recorded.

Acorn Woodpeckers mostly store acorns in their granaries, but they can store other nuts, too, including almonds, walnuts, hazelnuts, pecans, and even pinyon pine seeds. The woodpeckers prefer to harvest acorns from fewer than ten trees growing within about 500 feet (150 meters) of the granary tree, but they will venture much farther in search of acorns if necessary.

The granary trees would not be possible without a social group of cooperating Acorn Woodpeckers to build and defend them. I call these groups supersocial for the magnificent granaries that they and their descendants build and occupy over generations. But it turns out the groups are not always huge. They average a little more than four birds, though some groups are a lot larger. They have a complex breeding system with groups of one to eight males partnering with one to four females who lay their eggs communally in the same nest. There are also helpers who do not mate but instead help rear the young. The helpers are typically young from previous years.

In forty years of study at Berkeley's Hastings reservation, beginning in 1975, Koenig and his crew found that 41 percent of groups had just one breeding male and one breeding female. Half the groups had two or more breeding males, and a quarter of the groups had two or more breeding females. Only 16 percent of the groups had multiple male and female breeders. In addition to the breeders, half of all groups had nonbreeding helpers, with slightly more of them male than female.

Among Acorn Woodpeckers, when there are multiple males, they tend to be related to their co-breeders as brothers, half brothers, father and son, and so on. When there are multiple breeding females, they also tend to be related to one another, mostly as sisters or half sisters, but there can also be mother-daughter pairs and a few apparently unrelated females.[25]

This made me wonder if co-breeding Acorn Woodpeckers would show the same kind of reproductive competition shown by communal anis, Taiwan Yuhinas, and Pukekos. Since communal females are often related, I would expect less aggression and egg destruction. To study egg destruction, Koenig or someone on his team climbed up to the nests to evaluate the eggs. They could tell who laid each egg by monitoring when a female visited the nest and when a new egg appeared. They even learned the sizes and shapes of the eggs that were specific to a given female. They affirmed familial connections by DNA testing the young at twenty-one days old.

Acorn Woodpeckers sometimes lay an unusual kind of egg, a runt egg, but only when there are multiple layers in a nest. These eggs are tiny compared with regular eggs and do not have normal yolks, so they do not develop and the adult birds destroy them. Acorn Woodpeckers have more runt eggs than any other species—over 4 percent of all eggs are runt eggs, when no other species examined had over 1.5 percent, with 0.5 percent being the average.[26] The runt eggs tend to be the first-laid eggs in nests with more than one female laying. They are probably best known from studies of chickens.

Runt eggs were not the only ones destroyed in communal nests. Of 195 normal eggs laid in nineteen nesting attempts with three laying females, 32 percent of the eggs were destroyed before incubation started, and 5 percent more were destroyed after the eggs were incubated. Nearly all the destroyers were females. As happened with the anis, the earlier the eggs were laid, the more likely they were to be destroyed.

It is interesting that relatedness among the communal females did not seem to dampen their reproductive competition, though all females tended to successfully lay eggs.

Males within a group also compete for mating rights. Again, I expected the competition might be muted because they are generally related to one another, but that does not seem to be the case. Walt Koenig and his team looked into this in a paper with the fetching title "Are You My Baby?"[27]

Koenig and Ron Mumme counted all the acorns in the granary trees of twenty-four groups over seven years and found that groups had one thousand to two thousand acorns, with one hundred or two hundred acorns per bird.[28] The birds eat all the acorns until they are gone, by which time the next year's crop is maturing on the oak trees.

Koenig and his team looked at the impact of acorn abundance on Acorn Woodpeckers, assessing the former by counting acorns still in trees and the latter by looking at woodpecker survival in years with different numbers of acorns. In his years of counting the acorn crop, Koenig found that there were eight poor years, fourteen very good years, and twenty-one average years. In a good year, 90 percent of adult Acorn Woodpeckers survived and nearly that percentage of juveniles, whereas in a low-acorn year as few as 65 percent of adults and only 28 percent of juveniles survived. Clearly the juveniles take the biggest hit when there are fewer acorns.

Another way of looking at Acorn Woodpeckers' dependence on acorns is by the relative distribution of oaks and Acorn Woodpeckers. Walt Koenig and Joseph Haydock did just that using data from Christmas Bird Counts.[29] These counts go back to the early 1900s, when they were proposed as an alternative to a traditional Christmas hunt. They can last all day and involve a count circle with a 15-mile (24-kilometer) diameter. I have counted birds at Christmas Bird Counts in Texas, Michigan, and Missouri, and they are a lot of fun, though one always feels many birds are missed. Still, we do our best.

Koenig and Haydock used Christmas Bird Count data from thirty years, winters from 1959–1960 to 1989–1990 (from when I was in first grade to the year my last child was born). They looked only at states where Acorn Woodpeckers are reported to occur and separated the range into two different regions, the Pacific Coast (California and Oregon) and the Southwest (Arizona, New Mexico, and Texas). For oak tree abundance and number of species, they used published floras and distributional maps.

Koenig and Haydock discovered that Acorn Woodpeckers occurred where there were at least two species of oaks and went up in number as the number of species increased at a site, up to their top category of seven or more species. So one species of oak in an area did not mean much to Acorn Woodpeckers. This is because it is not just abundance of acorns that matters but also how they vary across years.

This brings us to the question of masting. A mast year is a year when trees produce more acorns than usual. Different tree species are not necessarily synchronized when it comes to acorn production, so if there are two or more species of oak, the odds of a catastrophic year with few acorns goes down. Using the detailed information on acorn counts they had at Hastings, Koenig and his team figured that the risk of crop failure was 34 percent a year if there was only one species of oak, 14 percent if there were two species, and only 2 percent if there were five species. It is the bad years and not the good ones that we need to look at to understand what it takes for Acorn Woodpeckers to survive.

Thus far I have focused on the work of Walt Koenig and others to paint a picture of an acorn-dependent communal species with bands of brothers and sororities of sisters nesting together who nevertheless squabble among themselves and fiercely defend totemic granary trees from others. It is a system that has made these birds abundant along the Pacific Coast as long as there are two or more species of oaks.

But there is another Acorn Woodpecker story and it comes from New Mexico and work by Peter Stacey, Carl Bock, and David Ligon.[30] From 1975 to 1984 they studied a population in Water Canyon, New Mexico, above 6,500 feet (1,980 meters) elevation in the Magdalena Mountains south of Albuquerque. There, the territories were along the streams and rivers where most of the oaks grew and where there were also huge cottonwoods. The woodpeckers formed their granaries and nested in the cottonwoods. The groups were territorial only around their granary trees in the canyon bottoms, not in their foraging grounds up the slopes. Only two main species of oaks were in the area.

There was not as much available habitat, compared with Hastings in California, and the population was accordingly smaller. Stacey and Ligon divided the territories into three categories according to how many acorns their granaries could hold. Small-granary territories had fewer than a thousand storage holes; medium-granary territories had one thousand to three thousand storage holes; and large-granary territories had more than three thousand storage holes. Simple pairs tended to be the breeders in this population. The probability of helpers varied with granary size. As many as 27 percent of yearlings became helpers from large-granary territories; 15 percent of yearlings became helpers from medium-granary territories; and only 2 percent of yearlings became helpers from small-granary territories. Small-granary groups were less likely to claim the territory the following year, because the group fell apart or moved to another territory. Only in years of plentiful acorns did any of the birds maintain territories from one year to the next.

During the study, the researchers banded 274 birds, 146 of which were young. Then they looked for the birds the following years. Most young had moved to a neighboring territory. It is unlikely they dispersed out of the area entirely, because the appropriate habitat was in small mountain ranges isolated by large stretches of grasslands. For their study, Stacey and Ligon assumed that birds they could not find had perished, likely a reasonable assumption. Hatchling survival at three years was about 20 percent for birds from large-granary territories and about 3 percent from medium-granary territories, and none survived from the small-granary territories. Yearling survival at five years was 20 percent, 3 percent, and 0 percent, respectively.

Ligon and Stacey returned to their study site in 1994 and 1995, a decade after the end of their earlier study.[31] Sadly, they found that the twenty-one territories of the original study had mostly lost their birds. In 1994, seven of the twenty-one territories had birds, and the following year only one did. This was because nearly all the old trees that served as granaries had died. The time period during which middle-age trees might have grown had been one of intensive cattle grazing, which prevented tree growth, so there were no trees of that age. The younger trees were too young to be good for granaries.

The Acorn Woodpeckers in California present an unusually complete story of communal sociality as a result of acorns (a resource) and granary trees (the ability to store that resource). The birds at Hastings had both the necessary granary trees and multiple oak species to buffer those years when one or another species failed to produce acorns in number. As a result, they had a complex breeding system with communal breeders and helpers. Water Canyon lacked the resources and had fewer oak species, and therefore birds typically bred in pairs with few or no helpers, effectively losing the supersociality of the California birds.

The next bird I will discuss is the White-winged Chough of Australia. Robert Heinsohn did his Ph.D. work on the social behavior of White-winged Choughs, bringing the story of this fascinating cooperative bird to the world.[32] Since then he has worked mostly on birds of great conservation interest in remote areas, and he is based at the Australian National University in Canberra.

White-winged Choughs superficially look a lot like crows. They are black with white patches on their wings and sinister-looking red eyes. They live in groups whose members are much more dependent on one another than crows are. Like crows, White-winged Choughs are in the Corvides, a group that has about eight hundred species.[33] In fact, within the group, White-winged Choughs, in the family Corcoracidae, are not at all closely related to the crows, which are in the family Corvidae. Instead, their closest relatives are a poorly known New Guinea family called the Melampittidae, which has only two species, and the Paradisaeidae, the birds-of-paradise.

White-winged Choughs are the most social I can imagine. They are always in groups of four to twenty birds, with each group usually made up of two breeding birds and their one- to four-year-old progeny.[34] These birds are completely dependent on helpers because they live in eucalyptus woodlands where food in the form of arthropods is scarce and hard to discover. The worms and scarab beetle larvae that are so important to them live in the leaf litter and in the ground. It takes a long time for young birds to learn how to find food that might require them to dig holes as deep as 5 inches (12.7 centimeters). Knowing where to dig and when to move on are skills that take time to acquire, Rob Heinsohn told me. This also means that nestlings need multiple adults to bring enough food for them. They do not defend their large feeding territories. They do, however, defend their nests and typically even keep one or two sentries there.[35] These sentries can be important because neighboring groups of choughs sometimes fly in and destroy the nests. Evidently, even though choughs cannot economically defend their food against competitors, they can try to rub out the competition.

White-winged Choughs rarely breed until they are five years old, even though they are sexually mature at three. Both male and female youngsters help the breeding adults rear young. They assist in all aspects of brood care, including incubation.[36] In small groups of five or fewer birds, young helpers less than a year old incubate as much as the breeders do, while in larger groups they incubate less, and in groups of more than twelve birds they incubate very little. Older helpers incubate equivalently in groups of all sizes. It is clear that incubation has a cost for one-year-old helpers, since in small groups they weigh less after incubation than before whereas in larger groups, where they incubate less, there is no weight loss.

Heinsohn found that no nests with fewer than two breeders and five helpers successfully reared young through the first winter.[37] Helpers were so essential that groups sometimes raided their neighbors and kidnapped recently fledged young to add to their own group. Battling adults lined up near neighbors and performed a dance Heinsohn calls the "wing-wave-tail-wag" (WWTW) display, in which they show the whites of their wings, engorge their eyes to make them even redder, and call loudly until one group leaves. During these battles, still-dependent young can end up in the opposing group. Heinsohn describes in detail a case of a group of fifteen that harassed a neighboring group of four that had recently fledged two young:

> *"Upon meeting, both groups flew into a tree, the smaller group leaving its young on the ground. While the other choughs were involved with WWTW displays about 8 meters above the ground, three helpers from the larger group flew to the opposing group's young on the ground and used WWTW displays which the eldest of the two followed. The small group was chased to another tree, and made no apparent attempt to retrieve their missing young."*

Display battles continued, and ultimately the other young also transferred to the larger group. Heinsohn had other stories of groups kidnapping or enticing young with mixed results. Some stayed, others returned home. Those that ended up in a new group were fed and cared for by their adoptive group and became permanent members, perhaps a better outcome than what awaited them in their small natal groups. These new helpers helped unrelated young but had a chance to become breeders in future years.

Another approach to studying the importance of helpers is to change the birds' access to food, as was done with the Sociable Weavers.[38] Christopher Boland from the Heinsohn team manipulated food availability in thirteen experimental groups and compared them with thirteen unmanipulated control nests. The groups varied in size, with between four and eleven birds, and as many control and experimental groups as possible were paired by group size and when they initiated nest building. The experimental nests got food morning and afternoon on a tray close to the base of the nesting tree. They received 2.2 pounds (1 kilogram) of shredded cheese mixed with an amino acid powder formulated for birds.

The researchers observed 1,918 nest visits in which 3,063 food items were delivered to the chicks. A given bird could eat as many as thirty pieces of cheese before hunting for arthropods, but they tended not to bring the cheese to the nestlings. Nestlings in the supplemented groups were fed four times as much food by the adults and helpers as those in the unsupplemented groups, and were much less likely to perish. By the time they were thirty days old, nearly 90 percent of chicks in the supplemented groups were still alive, though only a little more than half the chicks in the unsupplemented groups were.

Other interesting patterns reveal how White-winged Choughs have evolved to deal with scarcity. The eggs are laid at one-day intervals, and incubation begins with the first-laid egg, which therefore hatches first. This staggered hatching means that a natural dominance hierarchy is set up among the four or so young in the nest. If conditions are harsh, the last-hatched chick ends up perishing. If they are more benign, all might survive. In the experiment, the adults and helpers at unsupplemented nests preferentially fed the heavier first-hatched nestlings, resulting in the abovementioned pattern of nearly half the young dying. By contrast, the adults and helpers at supplemented nests, realizing there was sufficient food for all, fed the lighter chicks more than the heavier chicks, contributing to the survival of nearly all. Asynchronous hatching and adjusting care according to food availability are important for a bird living under highly variable food conditions, as this experiment revealed beautifully.

Just because a nestling has successfully fledged, survived a winter, and become a first-year helper does not mean it is in good shape. These year-old birds strive to learn how to find food in the forbidding environment. To do so they have a lot of choices to make. They need to do the work of feeding nestlings in order to be accepted in the group, but they also have to feed themselves. As Heinsohn discovered, they feed themselves before feeding anyone else, and they get better and better at finding food as they grow from one to five years old. This means that helpers can feed the nestlings more and more as they get older, but they still feed nestlings much less than they feed themselves. In larger groups of twelve or so, the nestlings get fed less per bird than in smaller groups, which means there is more pressure on individual helpers to do more feeding in smaller groups.

Young birds have a trick for staying in good standing with the group and still getting the food they need.[39] Chough nests are always attended by at least one group member who acts as a sentinel or broods the nestlings when they are small. Heinsohn explained that turnover happens when the bird at the nest stands on its rim as a helper approaches, and then watches as the arriving bird lands on the opposite nest rim and feeds the young, often with much excitement and display. Then the sentinel moves away, either remaining close as a guard or joining the foraging group farther away. The just-arrived bird starts guarding or preening the nestlings. It almost seems as if the sentinel is there to monitor the arriving bird and its food delivery. But Boland and his team saw helpers arrive with food in their beaks and stick them into the gaping mouths of a nestling, but rather than release the food, they kept it for themselves; to fool their predecessor at the nest, the deceiver kept their beak in the nestling's mouth until the other bird left. The researchers saw this happen 89 times along with visible swallowing motions by the helper as evidence they were keeping the food for themselves. There were 1,824 other times when the nestlings were actually fed.

White-winged Choughs qualify as supersocial birds who are always in groups whether foraging or caring for the young. The last bird in this chapter is also extremely social but lacks a heavy role for helpers.

The Pinyon Jay, the bird that introduced the chapter, lacks the crest of a Blue Jay but has a strong blue crown, darker cheeks of lapis blue, and a dusty-looking pale blue body. Females are only slightly lighter in color than the males. The Pinyon Jay is one of my favorite birds, and my friend Kris Johnson has dedicated herself to them. Johnson

left her faculty position at Rice University to move west to the University of New Mexico in Albuquerque, where she could do the research necessary to help preserve these iconic birds of the pinyon-juniper lands. You are lucky if you live near Pinyon Jays, as they are restricted to western North America, wherever there are many acres of pinyon pines.[40]

There is a reason Pinyon Jays take the name of the pinyon pine. The two need each other. The tree waits for the jays to disperse its seeds. These seeds have evolved into food, since they have thin shells, protrude from the cones that house them, and have no way of propelling themselves far from the cones and into the soil where they might grow. Seeds that fall from the cones under the mother tree may be cached by chipmunks, woodrats, and other birds, but these caches are too local to spread the seeds to new areas. Instead, pinyon seeds depend on Pinyon Jays to pull the seeds from the cones, carry them far away, and then plant them in an open area where the seeds might grow. This is not why the jays bury seeds in the ground; they are caching them for later harvest, only incidentally leaving some seeds behind that might germinate. Pinyon Jays have no contract with the trees that they will eat only so many seeds, just as they have no contract as to where they might fly to plant the seeds. Instead, the trees have evolved to adjust to the Pinyon Jays, who are presumably happy to eat all the seeds, but incapable of remembering every single cache. The trees can thrive even if only a tiny number of their seeds take root. Kris Johnson says that of all the birds and mammals that eat pinyon seeds, only the large flocks of Pinyon Jays can carry seeds far away from the source and cache them in ways that can replant an entire woodland decimated by insects, drought, or cutting.

To understand Pinyon Jays is to understand pinyon pines. Pinyon Jays generally associate with the trees *Pinus edulis* and *Pinus monophylla* that grow along with about three species of juniper, though in some places the birds eat and cache other pine seeds. Humans also treasure the delicious pine nuts.

The relationship between Pinyon Jays and pinyon pines is mutualistic, with one species flourishing only in the presence of the other. The bond between bird and tree goes back about thirty million years to the Tertiary Period, when the climate became hotter and drier on the Mexican and Colorado Plateaus and southwestern United States.[41]

But why do Pinyon Jays deserve a place in this chapter of supersocial birds? They do not have huge nest structures like the Sociable Weavers, or granary trees like the Acorn Woodpeckers. They do not have extensive groups of helpers so dependent on the group that they sometimes only pretend to feed the nestlings, like the White-winged Choughs. They are an anchor in this chapter and this book because they are social at so many levels—feeding in flocks, caching their pinyon seeds together, nesting close to one another, grouping their young after they leave the nest. If you see one Pinyon Jay, you will soon see the whole flock.

I like to think of a Pinyon Jay flock as an organism in itself, a velvet blue one that feathers the pine-green western landscape and enlivens it with their calls. The flock might be viewed as an organism because its parts, the individual birds, do not function without the whole. A single Pinyon Jay pair will not thrive. The home range that this organism occupies is huge.

Johnson and her team looked at exactly how much of the landscape a Pinyon Jay group needs to use. Their home range requires thousands of acres containing variable vegetation and plenty of pinyon pines. Within that home range is the nesting area, which requires nesting trees for up to around thirty pairs and covers as many as 100 acres (40 hectares). This area has to be a woodland, not just scattered trees. Within the nesting colony, an individual pair will choose a nesting tree and an area around it of an acre or two (0.4 to 0.8 hectare).

Why should a small bird weighing only around 3.5 ounces (100 grams) need thousands of acres to survive? The answer comes from the pinyon trees. Their distribution across the landscape is variable, sometimes dense in pinyon-juniper woodlands and sometimes more scattered with sagebrush and grasses intervening. And like the oaks we encountered with the Acorn Woodpeckers, pinyon pines produce mast crops, which means many more seeds in some years than others. However, the birds need seeds to eat and feed their young every year. This is the same problem that Acorn Woodpeckers, also dependent on a tree mast crop, have. Acorn Woodpeckers thrive only where there are multiple species of oak that are not synchronized in acorn production, so they do not need to roam far, a good thing, too, since their granary trees are fixed in space.

Pinyon Jays have a different solution, one that needs vast home ranges—if the local trees are not producing enough acorns, they can fly far enough to find other trees that are. Sometimes there are no trees producing seeds in the entire home range of a Pinyon Jay flock, forcing the birds to turn to other foods, especially insects, and sometimes forgoing reproducing. Johnson looked at exactly how large of an area Pinyon Jay flocks use at Kirtland Air Force Base, southeast of Albuquerque, New Mexico.[42] First, she located the nesting areas, and then she followed the birds to see where they foraged. She banded some of them and attached antenna radio transmitters to fourteen birds so that they could be tracked. One flock covered 17 square miles (44 square kilometers) in the breeding season, expanding to 22 square miles (57 square kilometers) for the whole year. Other studies have also found very large home ranges.[43]

The ranges included woodlands with pinyon pines and junipers but also savannas and grasslands. The more-open areas were used for caching pinyon pine seeds. At Kirtland Air Force Base, Johnson observed birds retrieving seeds from caches in two areas, the top of a recently burned hill and a south-facing grassy slope with scattered juniper trees.

The family lives of these wide-ranging birds can be challenging to investigate. Bird feeders can anchor even Pinyon Jays to a convenient place for study. We know the behaviors of these birds best from two flocks in Flagstaff, Arizona.[44] For Russell Balda, Pinyon Jay research started when a colleague, Gary Bateman, came to his office in 1968 to tell him of a flock near his rural neighborhood in Doney Park, northeast of Flagstaff. Balda visited Bateman's ranch and saw Pinyon Jays feeding with the chickens, then found their nests in the area. Later he discovered that two Flagstaff women, Gene Foster and Katharine Bartlett, had an extensive feeding station right at their home that had been visited by Pinyon Jays since the early 1960s. Foster and Bartlett had converted their garden to feeders and provided fifteen different kinds of enticing seeds and mealworms. Balda says the birds arrived in town as refugees from destruction of nearby pinyon-juniper woodlands, which displaced and ultimately killed thousands of Pinyon Jays. These birds formed a flock Balda called the Town Flock.

The flock member numbers change with the seasons, varying from a low of about a hundred in the early breeding season to a high of nearly two hundred in the fall, when the flock contains a lot of juveniles. Within the flocks, pairs and sometimes helpers form the family unit. Pairing up is serious for all birds but particularly so for Pinyon Jays since they mate for life. Courtship is intense and involves nuptial gifts by males to females who might crouch and quiver their wings as they receive the precious food. Courtship is most frequent among the youngsters that don't yet have a mate, though they do not generally breed until they are two years old. The youngsters that are competing for mates do so largely among those they grew up with, though they may be excited about females that have immigrated into the flock.

One courting youngster might try several potential partners, offering the same seed to four females in a row, for example. But even long-standing Pinyon Jay couples court each other, affirming their bonds, just as long-married humans might still have what has come to be called date night.

By late December courting intensifies. I love that when courting becomes serious, the pair leaves the flock to court in private. As John Marzluff and Balda put it: "Females crouch before their males with head and bill pointed upward at a slight angle and bill opened wide. Their wings are flapped and fluttered vigorously, as the female vocalizes loudly with a variety of loud kaws, chirrs, and soft musical notes."[45]

The research done by Balda and his students can put numbers on Pinyon Jays pairing for life. Of 104 pair-bond transitions, only 3 were cases of divorce. More common was that the male or the female died. When their mate dies, the remaining partner has the difficult task of finding a new mate. This is hard because most of the available mates are much younger, from the recent crop of yearlings. While it is true that when a bird loses its mate, it does its best to find a new one, Pinyon Jays do not leave a mate just because they are unsuccessful. Thirty-eight pairs who repeatedly failed to produce young stayed together in spite of their lack of success.

We can get an idea of what exactly the birds are looking for in a mate from work Kris Johnson did in an aviary.[46] The normal procedure would be to put a female with a choice of two males in a mate-choice arena, but Pinyon Jays cannot bear to be nearly alone and would call plaintively for others if put in that situation. Instead, Johnson housed a whole group of noisy birds next to the mate-choice arena so that a female could choose between two males without fearing abandonment. Johnson found that females preferred brighter-colored males but did not care about the male's size. She found that males were less picky about color but preferred dominant, larger females. Being larger is an advantage, as females have to incubate eggs through snowy nights, though small can be good for males that fly long distances to recover cached pinyon seeds.

Once Pinyon Jays have a mate, they get down to the serious business of breeding. They build large stick nests with thick insulation on the bottom.[47] They can nest earlier than many other birds because they have the cached pinyon seeds to rely on, though the young themselves will need to be fed insects. It is not unusual to see a female Pinyon Jay sitting still on a nest warming her eggs while she is covered in snow; this is why the nest needs such thick insulation. There are trade-offs on nest location, with higher nests shaking off their snow earlier than lower ones, while lower ones are less visible to crow and raven predators. If the birds nested high and the nest contents were taken by a predator, they would build a new nest in a lower and more concealed location.[48]

The most common number of eggs in a nest is four, with one laid per day. The female incubates the eggs starting with the third egg, so subsequent eggs will hatch later, setting them up for failure during poor years. Eggs hatch after about seventeen days, and then the nestlings stay in the nest for about three weeks before fledging and joining the other youngsters from the colony in a grouping crèche. The crèche is simply a group of twenty to sixty youngsters that stay fairly close together.

In the crèche the young are fed about every forty-five minutes during the day, and each feeding takes about four minutes. It is usually the parents that feed the chicks, but sometimes other birds do so. Balda found that of 313 feedings they observed, 13 percent were from nonparents. These feedings were not accidents, as birds clearly know their own young by the time they have left the nest. Balda's team did an experiment in which they briefly took the young away, put them in bags, and placed the bags at a feeder the parents visited that was nearly a mile from the nesting area. The young could hear and call for their parents, and the parents went straight to the correct bags, recognizing the calls of their own chicks.

Some of the other birds that fed the young were older siblings. Helpers are not a big part of Pinyon Jay sociality, but they do occur. Helpers tend to be smaller males that did not succeed in breeding themselves. The Town Flock team found that helpers assisted the parents with nestlings in various ways: guarding and grooming them, but most of all feeding them. A helper might bring the nestlings a grasshopper, pinyon seeds, or other food either alone or with one of the parents.

Balda and his team found that some families had helpers while others did not and that this pattern continued across generations. Of the extended families the researchers followed, about a fifth had helpers. The lineages with helpers tended to be more successful and to produce more sons than the lineages without helpers did. Interestingly, no helper lineages went extinct during the years of their study, while more than half of the non-helper lineages went extinct during that time. There were other differences. On average, the helper-lineage birds lived a year longer than birds in the non-helper lineages, largely due to differences in their first year: Nests with helpers had more surviving yearlings. The helper birds were more likely to become breeders in their natal flock or the nearby flock that the Balda team monitored. This makes me wonder why more families do not have helpers.

We cannot fully understand Pinyon Jay success without visiting their caching behavior in more detail. They collect the seeds, hide them by the thousands, and usually remember where they are months later, even under snow. The flocks of Pinyon Jays in Flagstaff had caching areas that were as big as 10 acres (4 hectares). Any given flock could have as many as seven such areas in their much larger home range.

So, how many seeds are they hiding in these areas? Marzluff and Balda estimated that each jay might be able to carry a maximum of fifty seeds per trip in its special enlarged throat. If it makes six trips a day, that is three hundred seeds a day. During a ninety-day caching season, that would amount to more than twenty-seven thousand seeds. Of course, in non-mast years, they may be storing many fewer seeds. Like everything else in their lives, Pinyon Jays both cache and harvest in groups.

It is hard to believe Pinyon Jays can find their cached seeds so precisely, so Balda and Alan Kamil did an experiment.[49] They made a board that had 180 holes about 2.2 inches (5.6 cm) in diameter that could be plugged with wood or left open, revealing a cup of sand in which the birds could bury or recover seeds. The room also had landmarks like rocks and logs, which the birds could use to orient themselves around the grid of holes. The researchers used Pinyon Jays they kept in captivity for a few months before the experiments were done, so they would be used to the artificial environment. They were allowed to explore the room and the sand holes, so they could see that these were good places for caching seeds.

Pinyon Jays are not comfortable caching unless they are in a group, though they do not want others to see exactly where they cache. But the researchers did not want the complexity of a group for the caching experiment. After all, the jays might not cache or recover cached seeds as well when other birds were watching them. They solved the problem by putting a cage of Pinyon Jays next to the experimental room, just as Johnson did in her mate-choice experiments. When the target jay heard the others, he calmed down and cached freely.

They let each bird cache some seeds, waited a week after caching, and put the bird back in the room. Did the jays remember where they cached seeds? They did. The seven Pinyon Jays they tested did much better than random at probing the sand traps where they had hidden seeds the week earlier.

Pinyon Jays were once common in the ancestral pinyon pine and juniper landscapes, but they are declining in number across their range. According to North American Breeding Bird Surveys, the declines over the last fifty years have been steep, especially in Nevada's Great Basin and west-central New Mexico, where Pinyon Jays were most common.[50] Overall, from 1970 to 2014, the Pinyon Jay population shrunk by 85 percent.

The losses are so concerning that Pinyon Jays are listed as vulnerable on the International Union for the Conservation of Nature's Red List of Threatened Species. The US Fish and Wildlife Service is currently considering a petition to list them as threatened or endangered under the Endangered Species Act.

Why are their numbers declining so much? Humans are largely to blame. In a 1981 book, Ronald Lanner estimated that between 400,000 and 525,000 acres (161,875 and 212,450 hectares) of pinyon pines were destroyed for charcoal and mining between 1859 and 1880, long before any Breeding Bird Surveys.[51] Other human endeavors continued the process, from pinto bean farming in southwestern Colorado to large-scale clearing for cattle pasture. The trees were knocked down wholesale through a process known as "chaining," where a chain is dragged between tractors, knocking down everything in its path. The US Forest Service argued that pinyon pines and juniper had invaded regions that were previously grasslands. Not true, according to Lanner and more recent studies.[52] Current woodland management practices are destroying foraging and nesting habitat. Land management agencies are taking advantage of money available for wildfire prevention by thinning out the pinyon-juniper woodlands.[53]

The best strategy to protect the Pinyon Jays would be large-scale landscape conservation, but if that is not feasible, smaller actions can help. Rather than thinning trees everywhere as humans interested in fire prevention tend to do, conserving the woodlands in which the jays nest is key. In one study comparing thinned, completely destroyed, and unthinned pinyon-juniper woodlands, Patrick Magee and his collaborators found that both thinning and completely destroying the vegetation resulted in fewer Pinyon Jays.[54] At Kirtland Air Force Base, a thinning ban on 15 percent of the study area would protect all the potential nesting habitat. Bans on human activities, military and recreational, in larger areas of pinyon pines could be limited to the few months in the fall when trees are producing their seeds, and even that could be limited to mast years in a given area. Humans can also help the birds by providing water sources. Feeders help, too, though they are no substitute for the birds' preferred pinyon pines. Johnson's group put out a feeder at Kirtland Air Force Base and found that the birds formed a nesting colony near it in the juniper woodland and savanna habitat, which is unusual for them to nest in.

Although these historical human impacts have certainly affected Pinyon Jays, the most important impact is happening now. It is the negative impact of climate change on pinyon pines and, consequently, on the birds. Andreas Wion and his collaborators looked at the role of dry years associated with a warming climate on pinyon pine masting.[55] They reviewed seed-cone production from 2004 to 2017 of 187 trees from twenty-eight sites along a 685-mile (1,100 kilometer) latitudinal gradient. There were fewer mast years in hotter, dryer areas. This is hurting Pinyon Jay populations.

Another study compared pinyon pine seed production between the 1970s and the 1980s at a series of sites throughout New Mexico and northwestern Oklahoma.[56] When the researchers revisited the same woodlands in the 1980s, they found a 40 percent decline in seed-cone production. The hotter it was in late summer, the fewer seed cones the pinyon pines produced.

A third study looked at a large-scale drought in 2002–2003 at sites with pinyon pine and juniper vegetation through Utah, Colorado, Arizona, and New Mexico and found widespread, large-scale mortality across much of the pinyon pine range.[57] In some places like Mesita del Buey in New Mexico and Mesa Verde in Colorado and sites near Flagstaff, Arizona, more than 90 percent of the pinyon pines died. Climate change is harming Pinyon Jays. We need more concerted efforts to preserve the trees they depend on in places where they can still thrive. Climate change impacts all birds, with dire consequences for some.

This chapter has covered four supersocial birds. Dry environments with variable resources characterize all four. All have found ways to survive, from the Sociable Weavers in the African Kalahari Desert to the White-winged Choughs of dry eucalyptus woodlands in Australia to the Acorn Woodpeckers and Pinyon Jays in the American West. Some rely on special structures, like the haystack nests of the Sociable Weavers and the granary trees of the Acorn Woodpeckers. Others take years to teach their young how to find food in hard ground. Some, like the White-winged Chough, are so dependent on large groups to feed their young, they will even kidnap nonrelative helpers. Pinyon Jays might seem quite independent because they breed in pairs, but each pair is entirely embedded in a large foraging and nesting group that is always together as the birds search for increasingly rare pinyon pines. In all four species, older young sometimes remain with their parents to raise their younger siblings, but this helping behavior is only part of what makes these birds supersocial. They also depend on one another, breeding in groups or colonies, and relying on each other for protection against predators, finding food, and sheltering from harsh environments. These are only a few examples of the extent to which some birds take social life.

We humans are particularly supersocial. We have many kinds of specialists who build our houses from materials earlier humans have refined, grow our food from plants long ago evolved for our use, provide our health care using methods discovered over time, and teach us what our ancestors have figured out. Together we also develop and use sophisticated and terrifying weapons against other human groups. Not only our sustenance and protection but also our joy comes from sharing our music, art, literature, dance, sports, and adventure. I could not have written this book without all the stories that ornithologists have shared. For better or worse, humans are the most supersocial organisms of all.

Conclusion

Why Are Birds So Social?

When I think back on the many birds this book has explored, some stand out. Common Redshanks on the east coast of England forage in groups along prey-rich mudflats, ever attentive to the sparrowhawks waiting in the bordering forests to turn one of them into a meal. Red-winged Blackbirds soar and settle in ever-larger groups as the American autumn progresses, feeding in spent wheat fields and soaring when an immature Bald Eagle clumsily approaches.

The roosts that most startled me were those of the Barn Swallows in Nigeria; birds swoop down by the thousands to settle in tall grasses where they are safer from natural predators but risk human harvest. And the Black Vultures in North Carolina roost together in long-used sites in trees or abandoned buildings, or on rusty old equipment; they are among the few species that exchange information on prey locations at the roost.

Chickadees and Tufted Titmice are the sentinels of North American mixed-species flocks, warning the shyer

Downy Woodpeckers and nuthatches. Fieldfares sometimes nest in denser colonies and can defeat predators with coordinated pooping on the intruder.

Seabirds must often fly or swim for hours to get to the best feeding grounds. No wonder they typically raise only one chick in a safe and often huge colony.

In Panama, Lance-tailed Manakins have beta males that dance for years to help the unrelated alpha achieve matings, often with the same females year after year. Satellite males in Ruffs increase the mating chances of dominants, but that is something the satellite will never become, as they are genetically different. The satellite's best option is a sneaky copulation.

Avian families are particularly complex. Baby birds generally need the care of at least both parents, particularly in songbirds like the Indigo Bunting, my spark bird. But this does not mean the breeders have identical interests, and philandering outside the pair bond is more the rule than the exception. Many species have helpers, which are usually the male young from a previous year. An exception is the Long-tailed Tit, who finds a relative to help only when independent breeding has not worked out.

Sociable Weavers in the African Kalahari Desert show familial cooperation with helpers in their nest chambers but also build enormous haystacks in camelthorn trees that require extended cooperation from unrelated birds for construction, making these birds supersocial.

Despite the considerable variation among these and other species, there are some commonalities. Nearly every species fears predation, and groups can protect against it. Group members may give alarm calls or overwhelm predators. Sometimes the groups only form incidentally as birds find places to nest that are safe from predators.

Relatives are generally more cooperative than nonrelatives, but nonrelatives sometimes do cooperate, even as they jostle for their own best interests. This begins with the mated pair, unrelated to each other though both are related to some of their nestlings, the mother typically more than the father. Many of the kinds of groups we have met are made up of unrelated individuals in flocks, roosts, colonies, leks, and communal nests. How these individuals try to achieve their selfish interests within these cooperative contexts is fascinating.

Grouping can take many different forms. These forms vary with the environment and the evolutionary history of the individual species, and discovering the role of each can lead to a deeper understanding of the birds' behaviors.

Each species has evolved to combat the threats present in the environment it finds itself in. As we change the environment in ever more profound ways, from turning wildness into farms and cities to altering the climate itself, we risk losing our birds. These losses begin small and then accelerate. Since I began my university studies at the University of Michigan in 1970, North American bird populations have been reduced by a third, a decline of more than three billion birds.[1] There are thousands of studies documenting the plight of the birds, studies worth reading. But reviewing them was not the purpose of this book. Here I tried to share my joy in the social lives of birds and the discoveries ornithologists have made about them. From the vantage point of understanding what the birds are doing, perhaps we can find it in our hearts to treasure them more.

As I researched all these birds and discovered their lives, I turned personally to the ornithologists who figured out what the birds are doing and why. And whether I already knew them or not, they were unfailingly generous with their time and their stories. I hope this book gives you a peek into their world as well as the birds' own.

Acknowledgments

This book would not be what it is without the help and collaboration of many people. From the beginning, my plan for the book was improved in countless discussions with my husband, David Queller. He read early drafts of every chapter and gave me insights on how to improve the narrative and its organization, and how to communicate what I most wanted to say.

My former graduate student Tyler Larsen also gave me comments at the early stages, offering thoughtful opinions on each chapter. Since Dave and Tyler both knew the conceptual material very well, they could suggest subjects that should have been there but were not. This book would be quite different without their help.

All the chapters had input by some of the scientists who contributed to the research I discuss. These scientists saved me all too often from embarrassing mistakes, though of course those that remain are my responsibility. Rob Cresswell discussed flocks with me and shared details of the Common Redshanks. Patty Parker brought the roosts of Black Vultures to life, and Guy Beauchamp shared the intricacies of his analyses of many species of roosting birds.

Doug Morse, Hari Sridhar, and Kimberly Sullivan educated me on various details and experiments on mixed-species flocks. I talked with Charles Brown about Cliff Swallow coloniality and John Hoogland about Bank Swallow coloniality. Doug Mock and Bob Montomerie kindly read the chapter on seabird colonies, making me wish I'd had more time to see northern colonies of gannets and puffins. David Westneat and Kris Johnson helped me understand Indigo Buntings and Great-tailed Grackles, respectively. Emily DuVal patiently teased apart the mysteries of Lance-tailed Manakins and their pairs of dancing males, and Dov Lank explained Ruffs and their satellite males. Gail Patricelli told me about Greater Sage-grouse and her marvelous robot. Helpers at the nest were particularly challenging to understand. Ron Mumme gave insights on Florida Scrub-jays while Ashleigh Griffin, Carlos Botero, and Andrew Cockburn helped me understand the role of environment in the evolution of helping. Sandy Vehrencamp, Jim Quinn, and Christie Riehl made Groove-billed Anis, Pukekos, and Greater Anis come alive for me in all their social complexity. The supersocial birds might have been the hardest of all had I not had so much help. Rita Covas and Gavin Leighton offered comments and explanation of Sociable Weavers. Walt Koenig helped me understand Acorn Woodpeckers and their environmental variability. Rob Heinsohn helped with White-winged Choughs. Finally, Kris Johnson and John Marzluff helped me understand Pinyon Jays.

I am deeply grateful to Anthony Bartley for his excellent drawings, which so well capture the social lives of birds.

I have been fortunate to have several stages of editors that have greatly improved the book's clarity. Maddy Frank of the Biology Department at Washington University in St. Louis went over every chapter. She was the first overall editor and so was able to catch the most egregiously bad writing, which allowed the subsequent editors to focus on lifting my writing a bit higher.

My daughter suggested I recruit Letta Page of Page-Smithing to give a final read of the text and to fix what needed fixing. Letta did this and more, making the text flow much more smoothly. There are so many ways to say things and some are better than others. Letta could find the better ways. She also taught me to vary my language and find new adjectives when they were needed, or drop extraneous ones. As an outsider to the field, she let me know when I was unclear. Even as I write this, I wonder how Letta might express my thanks more effectively. I feel better about writing just knowing that Letta is there to save me from my most awkward prose.

My editor at Tarcher, Tracy Behar, saved me from my too-technical self. She read the book carefully, fixing remaining awkwardnesses. She let me know when she thought the reader would not understand. She helped me cut tangents only an academic could love. I hope that what remains is crisper and more to the point. Tracy has also been an invaluable guide through the whole complex process of writing this book. Liz Koyfman helped in many ways with the production of this book, helping me understand what was needed at every stage. Maureen Klier copyedited this book, making me realize how essential her fixes were. I especially appreciated the all-too-many places that she was able to clarify the facts on which these stories depend.

My agent, Michelle Tessler, worked hard to find an excellent home for this book, and offered support throughout the process. She encouraged me to write this book, which I feared was too ambitious. She answered my questions quickly and carefully, giving me the confidence to plunge right into writing another book on birds. Michelle is simply wonderful.

The motivation for this book goes back to my early undergraduate days. Ever since I knocked on Richard D. Alexander's door in the Natural History Museum at the University of Michigan one autumn day in 1972 and asked if I could study in his group, the topic I have most loved has been social behavior. My own research into social interactions has been on *Polistes* wasps, stingless bees, and social amoebae, but I never lost interest in sociality in general. When I decided to write a book on social behavior, it seemed natural to turn to a group of organisms that was not only well studied but also already had a place in the hearts of many readers.

Throughout this project, my family has been my frame, my nest, my biggest support and source of comfort. My husband did much of the hard work of seeing me through, but the rest of the family helped, too. My children, Anna Mueller, Daniel Mueller, and Philip Queller, had no doubt that I could do this and encouraged me even when it meant missing summer hikes in northern Michigan, kayaking down the Crystal River, and camping trips. They also took up binoculars and searched for the birds with me when we did get out.

My parents brought me to nature in Michigan, in Europe, and in Mexico, and bought me my first field guides. They showed me how to hike, how to enjoy wild berries, and how to find and identify birds. This book is dedicated to them.

Notes

Introduction: Are Birds Social?

1. K. V. Rosenberg et al., "Decline of the North American Avifauna," *Science* 366 (2019): 120–124.

Chapter 1: Flocks: Can Many Do Better Than Few?

1. W. D. Hamilton, "Geometry for the Selfish Herd," *Journal of Theoretical Biology* 31, no. 2 (1971): 295–311.
2. J. van Gils, P. Wiersma, and G. M. Kirwan, "Common Redshank (*Tringa totanus*)," version 1.0, in *Birds of the World*, ed. J. del Hoyo et al. (Cornell Lab of Ornithology, 2020).
3. W. Cresswell, "Flocking Is an Effective Anti-predation Strategy in Redshanks, *Tringa totanus*," *Animal Behaviour* 47, no. 2 (1994): 433–442.
4. P. I. Mitchell, I. Scott, and P. R. Evans, "Vulnerability to Severe Weather and Regulation of Body Mass of Icelandic and British Redshank *Tringa totanus*," *Journal of Avian Biology* 31, no. 4 (2003): 511–521.
5. Cresswell, "Flocking Is an Effective Anti-predation Strategy in Redshanks."
6. Cresswell, "Flocking Is an Effective Anti-predation Strategy in Redshanks."

7. J. L. Quinn and W. Cresswell, "Testing Domains of Danger in the Selfish Herd: Sparrowhawks Target Widely Spaced Redshanks in Flocks," *Proceedings of the Royal Society B* 273 (2006): 2521–2526.

8. A. Sansom et al., "Vigilance Benefits and Competition Costs in Groups: Do Individual Redshanks Gain an Overall Foraging Benefit?," *Animal Behaviour* 75, no. 6 (2008): 1869–1875.

9. Mitchell, Scott, and Evans, "Vulnerability to Severe Weather and Regulation of Body Mass."

10. L. R. Malpas et al., "Continued Declines of Redshank *Tringa totanus* Breeding on Saltmarsh in Great Britain: Is There a Solution to This Conservation Problem?," *Bird Study* 60, no. 3 (2013): 370–383.

11. E. Sharps et al., "Nest Trampling and Ground Nesting Birds: Quantifying Temporal and Spatial Overlap Between Cattle Activity and Breeding Redshank," *Ecology and Evolution* 7, no. 16 (2017): 6622–6633.

12. C. H. Tuite, "Population Size, Distribution and Biomass Density of the Lesser Flamingo in the Eastern Rift Valley, 1974–76," *Journal of Applied Ecology* 16, no. 3 (1979): 765–775.

13. G. Beauchamp, "Flocking in Birds Is Associated with Diet, Foraging Substrate, Timing of Activity, and Life History," *Behavioral Ecology and Sociobiology* 76, no. 6 (2022): 74.

14. G. Beauchamp, "Reduced Flocking by Birds on Islands with Relaxed Predation," *Proceedings of the Royal Society B* 271 (2004): 1039–1042.

15. D. W. E. Sankey et al., "Absence of 'Selfish Herd' Dynamics in Bird Flocks Under Threat," *Current Biology* 31, no. 14 (2021): 3192–3198.e7.

16. J. E. Herbert-Read et al., "Initiation and Spread of Escape Waves Within Animal Groups," *Royal Society Open Science* 2, no. 4 (2015): 140355.

17. J. R. Usherwood et al., "Flying in a Flock Comes at a Cost in Pigeons," *Nature* 474 (2011): 494–497.

18. H. Weimerskirch et al., "Energy Saving in Flight Formation," *Nature* 413 (2001): 697–698.

19. J. Janák, "Spotting the Akh: The Presence of the Northern Bald Ibis in Ancient Egypt and Its Early Decline," *Journal of the American Research Center in Egypt* 46 (2010): 17–31.

20. B. Voelkl et al., "Matching Times of Leading and Following Suggest Cooperation Through Direct Reciprocity During V-Formation Flight in Ibis," *Proceedings of the National Academy of Sciences* 112, no. 7 (2015): 2115–2120.

21. E. L. C. Shepard, "Energy Economy in Flight," *Current Biology* 32 (2022): R672–R675.

22. P. Kerlinger and S. A. Gauthreaux Jr., "Seasonal Timing, Geographic Distribution, and Flight Behavior of Broad-winged Hawks During Spring Migration in South Texas: A Radar and Visual Study," *The Auk* 102, no. 4 (1985): 735–743.

Chapter 2: Communal Roosts: Why Sleep Together?

1. S. L. Lima et al., "Sleeping Under the Risk of Predation," *Animal Behaviour* 70, no. 4 (2005): 723–736; N. C. Rattenborg and G. Ungurean, "The Evolution and Diversification of Sleep," *Trends in Ecology and Evolution* 38, no. 2 (2023): 156–170.

2. A. A. Allen, *The Red-winged Blackbird: A Study in the Ecology of a Cat-tail Marsh* (Linnaean Society of New York, 1914), 43–128.

3. K. Yasukawa and W. A. Searcy, "Red-winged Blackbird (*Agelaius phoeniceus*)," version 1.0, in *Birds of the World*, ed. P. G. Rodewald (Cornell Lab of Ornithology, 2020).

4. P. J. Weatherhead and D. J. Hoysak, "Dominance Structuring of a Red-winged Blackbird Roost," *The Auk* 101, no. 3 (1984): 551–555.

5. B. Meanley, "The Roosting Behavior of the Red-winged Blackbird in the Southern United States," *Wilson Bulletin* 77, no. 3 (1965): 217–228.

6. S. B. Chaplin, "The Energetic Significance of Huddling Behavior in Common Bushtits (*Psaltriparus minimus*)," *The Auk* 99, no. 3 (1982): 424–430.

7. B. J. Hatchwell et al., "Factors Influencing Overnight Loss of Body Mass in the Communal Roosts of a Social Bird," *Functional Ecology* 23, no. 2 (2009): 367–372.

8. A. McGowan et al., "Competing for Position in the Communal Roosts of Long-tailed Tits," *Animal Behaviour* 72, no. 5 (2006): 1035–1043.

9. R. A. Noske and G. M. Kirwan, "Varied Sittella (*Daphoenositta chrysoptera*)," version 1.0, in *Birds of the World*, ed. J. del Joyo et al. (Cornell Lab of Ornithology, 2020).

10. R. A. Noske, "Huddle-Roosting Behaviour of the Varied Sittella *Daphoenositta chrysoptera* in Relation to Social Status," *Emu-Austral Ornithology* 85, no. 3 (1985): 188–194.

11. R. G. Bijlsma and B. van den Brink, "A Barn Swallow *Hirundo rustica* Roost Under Attack: Timing and Risks in the Presence of African Hobbies *Falco cuvieri*," *Ardea* 93, no. 1 (2005): 37–48.

12. P. Ward and A. Zahavi, "The Importance of Certain Assemblages of Birds as 'Information-Centres' for Food-Finding," *Ibis* 115, no. 4 (1973): 517–534.

13. D. W. Mock, T. C. Lamey, and D. B. A. Thompson, "Falsifiability and the Information Centre Hypothesis," *Ornis Scandinavica* 19, no. 3 (1988): 231–248.

14. N. J. Buckley, et al., "Black Vulture (*Coragyps atratus*)," version 2.0, in *Birds of the World*, ed. P. G. Rodewald and B. K. Keeney (Cornell Lab of Ornithology, 2022).

15. M. P. Wallace, P. G. Parker, and S. A. Temple, "An Evaluation of Patagial Markers for Cathartid Vultures," *Journal of Field Ornithology* 51, no. 4 (1980): 309–314.

16. P. G. Parker, T. A. Waite, and M. D. Decker, "Kinship and Association in Communally Roosting Black Vultures," *Animal Behaviour* 49, no. 2 (1995): 395–401; P. P. Rabenold, "Recruitment to Food in Black Vultures: Evidence for Following from Communal Roosts," *Animal Behaviour* 35, no. 6 (1987): 1775–1785; P. P. Rabenold, "Roost Attendance and Aggression in Black Vultures," *The Auk* 104, no. 4 (1987): 647–653.

17. M. D. Decker et al., "Monogamy in Black Vultures: Genetic Evidence from DNA Fingerprinting," *Behavioral Ecology* 4, no. 1 (1993): 29–35.

18. Rabenold, "Recruitment to Food in Black Vultures."

19. Rabenold, "Recruitment to Food in Black Vultures."

20. Parker, Waite, and Decker, "Kinship and Association in Communally Roosting Black Vultures"; P. P. Rabenold, "Family Associations in Communally Roosting Black Vultures," *The Auk* 103, no. 1 (1986): 32–41.

21. D. A. Buehler, "Bald Eagle (*Haliaeetus leucocephalus*)," version 2.0, in *Birds of the World*, ed. P. G. Rodewald and S. G. Mlodinow (Cornell Lab of Ornithology, 2022).

22. G. P. Keister Jr. and R. G. Anthony, "Characteristics of Bald Eagle Communal Roosts in the Klamath Basin, Oregon and California," *Journal of Wildlife Management* 47, no. 4 (1983): 1072–1079.

23. A. A. Yackel Adams, S. K. Skagen, and R. L. Knight, "Functions of Perch Relocations in a Communal Night Roost of Wintering Bald Eagles," *Canadian Journal of Zoology* 78, no. 5 (2011): 809–816.

24. Buehler, "Bald Eagle (*Haliaeetus leucocephalus*)," in Rodewald and Mlodinow, *Birds of the World*.

25. G. Beauchamp, "The Evolution of Communal Roosting in Birds: Origin and Secondary Losses," *Behavioral Ecology* 10, no. 6 (1999): 675–687.

26. S. L. Brusatte, J. K. O'Connor, and E. D. Jarvis, "The Origin and Diversification of Birds," *Current Biology* 25, no. 19 (2015): R888–R898.

Chapter 3: Mixed-Species Flocks: Follow the Alarm Caller

1. H. W. Bates, *The Naturalist on the River Amazons* (J. M. Dent and Sons, 1910).

2. M. H. Moynihan, "The Organization and Probable Evolution of Some Mixed Species Flocks of Neotropical Birds," in *Smithsonian Miscellaneous Collections* (Smithsonian Institute, 1962): 143:1–140.

3. J. Gradwohl and R. Greenberg, "The Formation of Antwren Flocks on Barro Colorado Island, Panamá," *The Auk* 97, no. 2 (1980): 385–395.

4. C. A. Munn and J. W. Terborgh, "Multi-species Territoriality in Neotropical Foraging Flocks," *The Condor* 81, no. 4 (1979): 338–347.

5. J. M. Winterbottom, "On Woodland Bird Parties in Northern Rhodesia," *Ibis* 85, no. 4 (1943): 437–442.

6. D. H. Morse, "Ecological Aspects of Some Mixed-Species Foraging Flocks of Birds," *Ecological Monographs* 40, no. 1 (1970): 119–168.

7. H. Sridhar and K. Shanker, "Using Intra-flock Association Patterns to Understand Why Birds Participate in Mixed-Species Foraging Flocks in Terrestrial Habitats," *Behavioral Ecology and Sociobiology* 68, no. 2 (2014): 185–196.

8. J. Suhonen, "Risk of Predation and Foraging Sites of Individuals in Mixed-Species Tit Flocks," *Animal Behaviour* 45, no. 6 (1993): 1193–1198.

9. A. S. Dolby and T. C. Grubb Jr., "Social Context Affects Risk Taking by a Satellite Species in a Mixed-Species Foraging Group," *Behavioral Ecology* 11, no. 1 (2000): 110–114.

10. K. A. Sullivan, "Information Exploitation by Downy Woodpeckers in Mixed-Species Flocks," *Behaviour* 91, no. 4 (1984): 294–311; K. A. Sullivan, "Selective Alarm Calling by Downy Woodpeckers in Mixed-Species Flocks," *The Auk* 102, no. 1 (1985): 184–187.

11. A. E. Martínez et al., "Deconstructing the Landscape of Fear in Stable Multi-species Societies," *Ecology* 98, no. 9 (2017): 2447–2455.

12. A. E. Martínez et al., "Fear-based Niche Shifts in Neotropical Birds," *Ecology* 99, no. 6 (2018): 1338–1346.

13. M. Jullien and J.-M. Thiollay, "Multi-species Territoriality and Dynamic of Neotropical Forest Understorey Bird Flocks," *Journal of Animal Ecology* 67, no. 2 (1998): 227–252; M. Jullien and J. Clobert, "The Survival Value of Flocking in Neotropical Birds: Reality or Fiction?," *Ecology* 81, no. 12 (2000): 3416–3430.

14. A. E. Martínez and J. P. Gomez, "Are Mixed-Species Bird Flocks Stable Through Two Decades?," *American Naturalist* 181, no. 3 (2013): E53–E59.

15. A. E. Martínez et al., "The Structure and Organisation of an Amazonian Bird Community Remains Little Changed After Nearly Four Decades in Manu National Park," *Ecology Letters* 26, no. 2 (2023): 335–346.

16. J.-M. Thiollay, "Influence of Selective Logging on Bird Species Diversity in a Guianan Rain Forest," *Conservation Biology* 6, no. 1 (1992): 47–63.

17. F. Zou et al., "The Conservation Implications of Mixed-Species Flocking in Terrestrial Birds, a Globally-Distributed Species Interaction Network," *Biological Conservation* 224 (2018): 267–276.

18. W. Marthy and D. R. Farine, "The Potential Impacts of the Songbird Trade on Mixed-Species Flocking," *Biological Conservation* 222 (2018): 222–231.

19. R. O. Bierregaard Jr. et al., "The Biological Dynamics of Tropical Rainforest Fragments," *BioScience* 42, no. 11 (1992): 859–866.

20. K. Mokross et al., "Decay of Interspecific Avian Flock Networks Along a Disturbance Gradient in Amazonia," *Proceedings of the Royal Society B* 281 (2014): 20132599.

Chapter 4: Colonies: Safe Places near Food

1. J. Burger, "A Model for the Evolution of Mixed-Species Colonies of Ciconiiformes," *Quarterly Review of Biology* 56, no. 2 (1981): 143–167.

2. R. D. Alexander, "The Evolution of Social Behavior," *Annual Review of Ecology and Systematics* 5 (1974): 325–383.

3. R. Dawkins, *The Selfish Gene* (Oxford University Press, 1976).

4. J. L. Hoogland and P. W. Sherman, "Advantages and Disadvantages of Bank Swallow (*Riparia riparia*) Coloniality," *Ecological Monographs* 46, no. 1 (1976): 33–58.

5. J. Janovy, *Keith County Journal* (University of Nebraska Press, 1996).

6. C. R. Brown and M. B. Brown, *Coloniality in the Cliff Swallow: The Effect of Group Size on Social Behavior* (University of Chicago Press, 1996).

7. C. R. Brown et al., "Cliff Swallow (*Petrochelidon pyrrhonota*)," version 1.0, in *Birds of the World*, ed. P. G. Rodewald (Cornell Lab of Ornithology, 2020).

8. C. R. Brown et al., "Cliff Swallow (*Petrochelidon pyrrhonota*)," in Rodewald, *Birds of the World*.

9. C. R. Brown et al., "The Cost of Ectoparasitism in Cliff Swallows Declines over 35 Years," *Ecological Monographs* 91, no. 2 (2021): e01446.

10. C. R. Brown and M. B. Brown, "Where Has All the Road Kill Gone?," *Current Biology* 23, no. 6 (2013): R233–R234.

11. H. Löhrl, "Well-aimed Defecation in the Fieldfare *Turdus pilaris*," *Journal für Ornithologie* 124 (1983): 271–279.

12. V. Haas, "Colonial and Single Breeding in Fieldfares, *Turdus pilaris* L.: A Comparison of Nesting Success in Early and Late Broods," *Behavioral Ecology and Sociobiology* 16, no. 2 (1985): 119–124.

13. C. G. Wiklund, "Fieldfare (*Turdus pilaris*) Breeding Success in Relation to Colony Size, Nest Position and Association with Merlins (*Falco columbarius*)," *Behavioral Ecology and Sociobiology* 11, no. 3 (1982): 165–172.

14. Wiklund, "Fieldfare (*Turdus pilaris*) Breeding Success."

15. H. S. Horn, "The Adaptive Significance of Colonial Nesting in the Brewer's Blackbird (*Euphagus cyanocephalus*)," *Ecology* 49, no. 4 (1968): 682–694; H. S. Horn, "Social Behavior of Nesting Brewer's Blackbirds," *The Condor* 72, no. 1 (1970): 15–23.

16. J. L. Sachs et al., "Evolution of Coloniality in Birds: A Test of Hypotheses with the Red-necked Grebe (*Podiceps grisegena*)," *The Auk* 124, no. 2 (2007): 628–642; G. L. Nuechterlein et al., "Red-necked Grebes Become Semicolonial When Prime Nesting Substrate Is Available," *The Condor* 105, no. 1 (2003): 80–94.

17. B. E. Stout and G. L. Nuechterlein, "Red-necked Grebe (*Podiceps grisegena*)," version 1.0, in *Birds of the World*, ed. S. M. Billerman (Cornell Lab of Ornithology, 2020).

18. R. W. Allen and M. M. Nice, "A Study of the Breeding Biology of the Purple Martin (*Progne subis*)," *American Midland Naturalist* 47, no. 3 (1952): 606–665.

19. G. Nobles, *John James Audubon: The Nature of the American Woodsman* (University of Pennsylvania Press, 2017).

20. B. J. Stutchbury, "Coloniality and Breeding Biology of Purple Martins (*Progne subis hesperia*) in Saguaro Cacti," *The Condor* 93, no. 3 (1991): 666–675.

21. Allen and Nice, "A Study of the Breeding Biology of the Purple Martin."

22. E. S. Morton, L. Forman, and M. Braun, "Extrapair Fertilizations and the Evolution of Colonial Breeding in Purple Martins," *The Auk* 107, no. 2 (1990): 275–283.

Chapter 5: Seabird Colonies: How to Rear Young by the Largest Larder on Earth

1. E. A. Schreiber and J. Burger, eds., *Biology of Marine Birds* (CRC Press, 2001).

2. M. de L. Brooke, "The Food Consumption of the World's Seabirds," *Proceedings of the Royal Society B* 271 (2004): S246–S248.

3. M. Brooke, *The Manx Shearwater* (T. & A. D. Poyser, 1990).

4. D. L. Lack, *Ecological Adaptations for Breeding in Birds* (Methuen and Co., 1968).

5. C. Rolland, E. Danchin, and M. de Fraipont, "The Evolution of Coloniality in Birds in Relation to Food, Habitat, Predation, and Life-history Traits: A Comparative Analysis," *American Naturalist* 151, no. 6 (1998): 514–529.

6. D. J. Anderson and P. J. Hodum, "Predator Behavior Favors Clumped Nesting in an Oceanic Seabird," *Ecology* 74, no. 8 (1993): 2462–2464.

7. T. R. Birkhead, "The Effect of Habitat and Density on Breeding Success in the Common Guillemot (*Uria aalge*)," *Journal of Animal Ecology* 46, no. 3 (1977): 751–764.

8. D. N. Nettleship, "Breeding Success of the Common Puffin (*Fratercula arctica* L.) on Different Habitats at Great Island, Newfoundland," *Ecological Monographs* 42 (1972): 239–268.

9. J. Coulson, "Differences in the Quality of Birds Nesting in the Centre and on the Edges of a Colony," *Nature* 217 (1968): 478–479.

10. J. Coulson, "Colonial Breeding in Seabirds," in Schreiber and Burger, *Biology of Marine Birds*, 87–113.

11. A. J. Gaston, R. C. Ydenberg, and G. E. J. Smith, "Ashmole's Halo and Population Regulation in Seabirds," *Marine Ornithology* 35, no. 2 (2007): 119–126.

12. N. P. Ashmole, "The Regulation of Numbers of Tropical Oceanic Birds," *Ibis* 103b (1963): 458–473.

13. R. W. Furness and T. R. Birkhead, "Seabird Colony Distributions Suggest Competition for Food Supplies During the Breeding Season," *Nature* 311 (1984): 655–656.

14. V. L. Birt et al., "Ashmole's Halo: Direct Evidence for Prey Depletion by a Seabird," *Marine Ecology Progress Series* 40, no. 3 (1987): 205–208.

15. S. Lewis et al., "Evidence of Intra-specific Competition for Food in a Pelagic Seabird," *Nature* 412 (2001): 816–819.

16. H. Sandvik et al., "Modelled Drift Patterns of Fish Larvae Link Coastal Morphology to Seabird Colony Distribution," *Nature Communications* 7 (2016): 11599.

17. A. Ancel, M. Beaulieu, and C. Gilbert, "The Different Breeding Strategies of Penguins: A Review," *Comptes Rendus Biologies* 336, no. 1 (2013): 1–12.

18. D. G. Ainley, N. Nur, and E. J. Woehler, "Factors Affecting the Distribution and Size of Pygoscelid Penguin Colonies in the Antarctic," *The Auk* 112, no. 1 (1995): 171–182.

19. M. P. Dias et al., "Threats to Seabirds: A Global Assessment," *Biological Conservation* 237 (2019): 525–537.

20. D. W. Steadman, "Prehistoric Extinctions of Pacific Island Birds: Biodiversity Meets Zooarchaeology," *Science* 267 (1995): 1123–1131.

21. S. C. Votier et al., "An Overview of the Impacts of Fishing on Seabirds, Including Identifying Future Research Directions," *ICES Journal of Marine Science* 80, no. 9 (2023): 2380–2392.

22. Dias et al., "Threats to Seabirds: A Global Assessment."

23. F. Cusset et al., "Arctic Seabirds and Shrinking Sea Ice: Egg Analyses Reveal the Importance of Ice-derived Resources," *Scientific Reports* 9 (2019): 15405.

24. Steadman, "Prehistoric Extinctions of Pacific Island Birds."

25. Steadman, "Prehistoric Extinctions of Pacific Island Birds."

26. H. P. Jones et al., "Severity of the Effects of Invasive Rats on Seabirds: A Global Review," *Conservation Biology* 22, no. 1 (2008): 16–26.

27. J. F. Piatt et al., "Extreme Mortality and Reproductive Failure of Common Murres Resulting from the Northeast Pacific Marine Heatwave of 2014–2016," *PLoS One* 15, no. 1 (2020): e0226087.

28. J. Forcada and P. N. Trathan, "Penguin Responses to Climate Change in the Southern Ocean," *Global Change Biology* 15, no. 7 (2009): 1618–1630.

29. Forcada and Trathan, "Penguin Responses to Climate Change in the Southern Ocean."

Chapter 6: Leks: Where Males Dance and Females Choose

1. N. B. Davies, J. R. Krebs, and S. A. West, *An Introduction to Behavioural Ecology* (Wiley-Blackwell, 2012).

2. G. Borgia, "Sexual Selection and the Evolution of Mating Systems," in *Sexual Selection and Reproductive Competition in Insects*, ed. M. S. Blum and N. A. Blum (Academic Press, 1979), 19–80; M. Kirkpatrick and M. J. Ryan, "The Evolution of Mating Preferences and the Paradox of the Lek," *Nature* 350 (1991): 33–38.

3. Borgia, "Sexual Selection and the Evolution of Mating Systems"; Kirkpatrick and Ryan, "The Evolution of Mating Preferences and the Paradox of the Lek."

4. W. D. Hamilton and M. Zuk, "Heritable True Fitness and Bright Birds: A Role for Parasites?," *Science* 218 (1982): 384–387.

5. M. A. Schroeder, J. R. Young, and C. E. Braun, "Greater Sage-grouse (*Centrocercus urophasianus*)," version 1.0, in *Birds of the World*, ed. A. F. Poole and Frank B. Gill (Cornell Lab of Ornithology, 2020).

6. J. W. Scott, "Mating Behavior of the Sage Grouse," *The Auk* 59, no. 4 (1942): 477–498.

7. R. H. Wiley, "Territoriality and Non-random Mating in Sage Grouse, *Centrocercus urophasianus*," *Animal Behaviour Monographs* 6 (1973): 85–99.

8. R. M. Gibson and J. W. Bradbury, "Sexual Selection in Lekking Sage Grouse: Phenotypic Correlates of Male Mating Success," *Behavioral Ecology and Sociobiology* 18, no. 2 (1985): 117–123.

9. G. L. Patricelli and A. H. Krakauer, "Tactical Allocation of Effort Among Multiple Signals in Sage Grouse: An Experiment with a Robotic Female," *Behavioral Ecology* 21, no. 1 (2010): 97–106.

10. K. Semple, R. K. Wayne, and R. M. Gibson, "Microsatellite Analysis of Female Mating Behaviour in Lek-breeding Sage Grouse," *Molecular Ecology* 10, no. 8 (2001): 2043–2048.

11. D. B. Lank and J. Dale, "Visual Signals for Individual Identification: The Silent 'Song' of Ruffs," *The Auk* 118, no. 3 (2001): 759–765.

12. A. J. Hogan-Warburg, "Social Behavior of the Ruff, *Philomachus pugnax* (L.)" *Ardea* 54 (1966): 111–226.

13. D. B. Lank et al., "Genetic Polymorphism for Alternative Mating Behaviour in Lekking Male Ruff *Philomachus pugnax*," *Nature* 378 (1995): 59–62.

14. S. Lamichhaney et al., "Structural Genomic Changes Underlie Alternative Reproductive Strategies in the Ruff (*Philomachus pugnax*)," *Nature Genetics* 48 (2016): 84–88; C. Küpper et al., "A Supergene Determines Highly Divergent Male Reproductive Morphs in the Ruff," *Nature Genetics* 48 (2016): 79–83.

15. J. Jukema and T. Piersma, "Permanent Female Mimics in a Lekking Shorebird," *Biology Letters* 2, no. 2 (2006): 161–164.

16. J. Hill et al., "Low Mutation Load in a Supergene Underpinning Alternative Male Mating Strategies in Ruff (*Calidris pugnax*)," *Molecular Biology and Evolution* 40, no. 12 (2023): msad224.

17. J. D. M. Tolliver et al., "Fitness Benefits from Co-display Favour Subdominant Male–Male Partnerships Between Phenotypes," *Animal Behaviour* 197, no. 1 (2023): 131–154.

18. D. B. Lank et al., "High Frequency of Polyandry in a Lek Mating System," *Behavioral Ecology* 13, no. 2 (2002): 209–215.

19. E. H. DuVal, "Cooperative Display and Lekking Behavior of the Lance-tailed Manakin (*Chiroxiphia lanceolata*)," *The Auk* 124, no. 4 (2007): 1168–1185.

20. E. H. DuVal, "Adaptive Advantages of Cooperative Courtship for Subordinate Male Lance-tailed Manakins," *American Naturalist* 169, no. 4 (2007): 423–432.

21. E. H. DuVal and B. Kempenaers, "Sexual Selection in a Lekking Bird: The Relative Opportunity for Selection by Female Choice and Male Competition," *Proceedings of the Royal Society B* 275 (2008): 1995–2003.

22. P. R. Rivers and E. H. DuVal, "Multiple Paternity in a Lek Mating System: Females Mate Multiply When They Choose Inexperienced Sires," *Journal of Animal Ecology* 89, no. 5 (2020): 1142–1152.

23. E. H. DuVal, "Does Cooperation Increase Helpers' Later Success as Breeders? A Test of the Skills Hypothesis in the Cooperatively Displaying Lance-tailed Manakin," *Journal of Animal Ecology* 82, no. 4 (2013): 884–893.

24. R. J. Sardell, B. Kempenaers, and E. H. DuVal, "Female Mating Preferences and Offspring Survival: Testing Hypotheses on the Genetic Basis of Mate Choice in a Wild Lekking Bird," *Molecular Ecology* 23, no. 4 (2014): 933–946.

25. E. H. DuVal and J. A. Kapoor, "Causes and Consequences of Variation in Female Mate Search Investment in a Lekking Bird," *Behavioral Ecology* 26, no. 6 (2015): 1537–1547.

26. E. H. DuVal, "Female Mate Fidelity in a Lek Mating System and Its Implications for the Evolution of Cooperative Lekking Behavior," *American Naturalist* 181, no. 2 (2013): 213–222.

27. R. J. Sardell and E. H. DuVal, "Differential Allocation in a Lekking Bird: Females Lay Larger Eggs and Are More Likely to Have Male Chicks When They Mate with Less Related Males," *Proceedings of the Royal Society B* 281 (2014): 20132386.

Chapter 7: Mate Choice and Parental Care: Competition in the Family

1. D. J. Varricchio et al., "Avian Paternal Care Had Dinosaur Origin," *Science* 322 (2008): 1826–1828.

2. G. M. Kirwan, A. Korthals, and C. E. Hodes, "Greater Rhea (*Rhea americana*)," version 2.0, in *Birds of the World*, ed. B. K. Keeney (Cornell Lab of Ornithology, 2021).

3. G. J. Fernández and J. C. Reboreda, "Effects of Clutch Size and Timing of Breeding on Reproductive Success of Greater Rheas," *The Auk* 115, no. 2 (1998): 340–348.

4. D. N. Jones, R. W. R. J. Dekker, and C. S. Roselaar, *The Megapodes: Megapodiidae* (Oxford University Press, 1995).

5. Jones, Dekker, and Roselaar, *The Megapodes*; A. Elliott, G. M. Kirwan, and D. Christie, "Orange-footed Megapode (*Megapodius reinwardt*)," version 1.0, in *Birds of the World*, ed. J. del Hoyo et al. (Cornell Lab of Ornithology, 2020).

6. A. Elliott and G. M. Kirwan, "Australian Brushturkey (*Alectura lathami*)," version 1.0, in *Birds of the World*, ed. J. del Hoyo et al. (Cornell Lab of Ornithology, 2020).

7. A. Cockburn, "Prevalence of Different Modes of Parental Care in Birds," *Proceedings of the Royal Society B* 273 (2006): 1375–1383.

8. K. Johnson and B. D. Peer, "Great-tailed Grackle (*Quiscalus mexicanus*)," version 2.0, in *Birds of the World*, ed. P. G. Rodewald and B. K. Keeney (Cornell Lab of Ornithology, 2022); S. T. Emlen, P. H. Wrege, and M. S. Webster, "Cuckoldry as a Cost of Polyandry in the Sex-role-reversed Wattled Jacana, *Jacana jacana*," *Proceedings of the Royal Society B* 265 (1998): 2359–2364.

9. O. E. Bray, J. J. Kennelly, and J. L. Guarino, "Fertility of Eggs Produced on Territories of Vasectomized Red-winged Blackbirds," *Wilson Bulletin* 87, no. 2 (1975): 187–195.

10. P. J. Weatherhead and P. T. Boag, "Pair and Extra-pair Mating Success Relative to Male Quality in Red-winged Blackbirds," *Behavioral Ecology and Sociobiology* 37, no. 2 (1995): 81–91.

11. L. Brouwer and S. C. Griffith, "Extra-pair Paternity in Birds," *Molecular Ecology* 28, no. 22 (2019): 4864–4882.

12. R. B. Payne, "Indigo Bunting (*Passerina cyanea*)," version 1.0, in *Birds of the World*, ed. A. F. Poole (Cornell Lab of Ornithology, 2020).

13. D. F. Westneat, "Extra-pair Copulations in a Predominantly Monogamous Bird: Observations of Behaviour," *Animal Behaviour* 35, no. 3 (1987): 865–876.

14. A. Dyrcz, "Great Reed Warbler (*Acrocephalus arundinaceus*)," version 1.0, in *Birds of the World*, ed. J. del Hoyo et al. (Cornell Lab of Ornithology, 2020).

15. C. K. Catchpole, "Variation in the Song of the Great Reed Warbler *Acrocephalus arundinaceus* in Relation to Mate Attraction and Territorial Defence," *Animal Behaviour* 31, no. 4 (1983): 1217–1225.

16. C. K. Catchpole, "Song Repertoires and Reproductive Success in the Great Reed Warbler *Acrocephalus arundinaceus*," *Behavioral Ecology and Sociobiology* 19, no. 6 (1986): 439–445.

17. D. Hasselquist, S. Bensch, and T. von Schantz, "Correlation Between Male Song Repertoire, Extra-pair Paternity and Offspring Survival in the Great Reed Warbler," *Nature* 381 (1996): 229–232.

18. Hasselquist, Bensch, and von Schantz, "Correlation Between Male Song Repertoire."

19. Hasselquist, Bensch, and von Schantz, "Correlation Between Male Song Repertoire."

20. W. Wehtje, "The Range Expansion of the Great-tailed Grackle (*Quiscalus mexicanus Gmelin*) in North America Since 1880," *Journal of Biogeography* 30 (2003): 1593–1607.

21. K. Johnson et al., "Male Mating Strategies and the Mating System of Great-tailed Grackles," *Behavioral Ecology* 11, no. 2 (2000): 132–141.

22. A. Cockburn et al., "Superb Fairy-wrens: Making the Worst of a Good Job," in *Cooperative Breeding in Vertebrates: Studies of Ecology, Evolution, and Behavior*, ed. W. D. Koenig and J. L. Dickinson (Cambridge University Press, 2016), 133–149.

23. A. H. Dalziell and A. Cockburn, "Dawn Song in Superb Fairy-wrens: A Bird That Seeks Extrapair Copulations During the Dawn Chorus," *Animal Behaviour* 75, no. 2 (2008): 489–500; M. Double and A. Cockburn, "Pre-dawn Infidelity: Females Control Extra-pair Mating in Superb Fairy-wrens," *Proceedings of the Royal Society B* 267 (2000): 465–470.

24. P. O. Dunn and A. Cockburn, "Extrapair Mate Choice and Honest Signaling in Cooperatively Breeding Superb Fairy-wrens," *Evolution* 53, no. 3 (1999): 938–946.

25. S. Calhim et al., "Maintenance of Sperm Variation in a Highly Promiscuous Wild Bird," *PLoS One* 6, no. 12 (2011): e28809.

26. D. C. Queller, "Why Do Females Care More Than Males?," *Proceedings of the Royal Society B* 264 (1997): 1555–1557.

27. N. Burley, "Sexual Selection for Aesthetic Traits in Species with Biparental Care," *American Naturalist* 127, no. 4 (1986): 415–445.

28. D. Wang et al., "Irreproducible Text-book 'Knowledge': The Effects of Color Bands on Zebra Finch Fitness," *Evolution* 72, no. 4 (2018): 961–976.

29. B. C. Sheldon et al., "Ultraviolet Colour Variation Influences Blue Tit Sex Ratios," *Nature* 402 (1999): 874–877.

30. A. Johnsen et al., "Male Sexual Attractiveness and Parental Effort in Blue Tits: A Test of the Differential Allocation Hypothesis," *Animal Behaviour* 70, no. 4 (2005): 877–888.

31. M. Petrie and T. Halliday, "Experimental and Natural Changes in the Peacock's (*Pavo cristatus*) Train Can Affect Mating Success," *Behavioral Ecology and Sociobiology* 35, no. 3 (1994): 213–217.

32. C. Darwin et al., *The Correspondence of Charles Darwin: 1821–1860*, vol. 13 (Cambridge University Press, 2002).

33. M. Petrie and A. Williams, "Peahens Lay More Eggs for Peacocks with Larger Trains," *Proceedings of the Royal Society B* 251 (1993): 127–131.

34. M. Petrie, "Female Moorhens Compete for Small Fat Males," *Science* 220 (1983): 413–415.

35. D. Hanley, G. Heiber, and D. C. Dearborn, "Testing an Assumption of the Sexual-Signaling Hypothesis: Does Blue-green Egg Color Reflect Maternal Antioxidant Capacity," *The Condor* 110, no. 4 (2008): 767–771.

36. T. B. Ryder et al., "The Ecological–Evolutionary Interplay: Density-dependent Sexual Selection in a Migratory Songbird," *Ecology and Evolution* 2, no. 5 (2012): 976–987.

37. J. J. Soler et al., "Blue and Green Egg-color Intensity Is Associated with Parental Effort and Mating System in Passerines: Support for the Sexual Selection Hypothesis," *Evolution* 59, no. 3 (2005): 636–644.

38. W. D. Hamilton, "The Genetical Evolution of Social Behaviour," pts. 1 and 2, *Journal of Theoretical Biology* 7, no. 1 (1964): 1–52.

39. R. Dawkins, *The Selfish Gene* (Oxford University Press, 1976).

40. R. L. Trivers, "Parent–Offspring Conflict," *American Zoologist* 14, no. 1 (1974): 249–264.

41. R. Kilner, "Mouth Colour Is a Reliable Signal of Need in Begging Canary Nestlings," *Proceedings of the Royal Society B* 264 (1997): 963–968.

42. D. W. Mock, M. B. Dugas, and S. A. Strickler, "Honest Begging: Expanding from Signal of Need," *Behavioral Ecology* 22, no. 5 (2011): 909–917.

43. S. M. Leech and M. L. Leonard, "Begging and the Risk of Predation in Nestling Birds," *Behavioral Ecology* 8, no. 6 (1997): 644–646.

44. J. V. Briskie, C. T. Naugler, and S. M. Leech, "Begging Intensity of Nestling Birds Varies with Sibling Relatedness," *Proceedings of the Royal Society B* 258 (1994): 73–78.

45. T. M. Jones et al., "Parental Benefits and Offspring Costs Reflect Parent–Offspring Conflict over the Age of Fledging Among Songbirds," *Proceedings of the National Academy of Sciences* 117, no. 48 (2020): 30539–30546.

Chapter 8: Families with Helpers: Older Siblings, Lonely Bachelors, and More

1. G. E. Woolfenden and J. W. Fitzpatrick, *The Florida Scrub Jay: Demography of a Cooperative-Breeding Bird*, vol. 20 (Princeton University Press, 1984).

2. Woolfenden and Fitzpatrick, *The Florida Scrub Jay.*

3. R. L. Mumme, "Do Helpers Increase Reproductive Success? An Experimental Analysis in the Florida Scrub Jay," *Behavioral Ecology and Sociobiology* 31, no. 5 (1992): 319–328.

4. E. O. Wilson, *Sociobiology: The New Synthesis* (Belknap Press, 1975).

5. J. W. Fitzpatrick et al., "Florida Scrub-jays: Oversized Territories and Group Defense in a Fire-maintained Habitat," in *Cooperative Breeding in Vertebrates: Studies of Ecology, Evolution, and Behavior*, ed. W. D. Koenig and J. L. Dickinson (Cambridge University Press, 2016), 77–96.

6. D. R. Breininger and G. M. Carter, "Territory Quality Transitions and Source-Sink Dynamics in a Florida Scrub-jay Population," *Ecological Applications* 13, no. 2 (2003): 516–529.

7. A. Cockburn, "Evolution of Helping Behavior in Cooperatively Breeding Birds," *Annual Review of Ecology and Systematics* 29 (1998): 141–177.

8. H. A. Ford et al., "The Relationship Between Ecology and the Incidence of Cooperative Breeding in Australian Birds," *Behavioral Ecology and Sociobiology* 22, no. 4 (1988): 239–249.

9. N. Margraf and A. Cockburn, "Helping Behaviour and Parental Care in Fairy-wrens (*Malurus*)," *Emu-Austral Ornithology* 113, no. 3 (2013): 294–301.

10. A. Cockburn et al., "Superb Fairy-wrens: Making the Worst of a Good Job," in Koenig and Dickinson, *Cooperative Breeding in Vertebrates*, 133–149.

11. A. Peters, A. Cockburn, and R. Cunningham, "Testosterone Treatment Suppresses Paternal Care in Superb Fairy-wrens, *Malurus cyaneus*, Despite Their Concurrent Investment in Courtship," *Behavioral Ecology and Sociobiology* 51, no. 6 (2002): 538–547.

12. A. F. Russell et al., "Reduced Egg Investment Can Conceal Helper Effects in Cooperatively Breeding Birds," *Science* 317 (2007): 941–944.

13. A. Cockburn et al., "Can We Measure the Benefits of Help in Cooperatively Breeding Birds: The Case of Superb Fairy-wrens *Malurus cyaneus*?," *Journal of Animal Ecology* 77, no. 3 (2008): 430–438.

14. B. J. Hatchwell, "Long-tailed Tits: Ecological Causes and Fitness Consequences of Redirected Helping," in Koenig and Dickinson, *Cooperative Breeding in Vertebrates*, 39–57.

15. C. K. Cornwallis et al., "Cooperation Facilitates the Colonization of Harsh Environments," *Nature Ecology and Evolution* 1 (2017): 0057.

16. Cornwallis et al., "Cooperation Facilitates the Colonization of Harsh Environments."

17. M. Griesser et al., "Family Living Sets the Stage for Cooperative Breeding and Ecological Resilience in Birds," *PLOS Biology* 15, no. 6 (2017): e2000483.

18. A. S. Griffin and S. A. West, "Kin Discrimination and the Benefit of Helping in Cooperatively Breeding Vertebrates," *Science* 302 (2003): 634–636.

19. H.-U. Reyer, "Flexible Helper Structure as an Ecological Adaptation in the Pied Kingfisher (*Ceryle rudis rudis* L.)," *Behavioral Ecology and Sociobiology* 6, no. 3 (1980): 219–227; H.-U. Reyer, "Investment and Relatedness: A Cost/Benefit Analysis of Breeding and Helping in the Pied Kingfisher (*Ceryle rudis*)," *Animal Behaviour* 32, no. 4 (1984): 1163–1178.

20. S. Legge, "Helper Contributions in the Cooperatively Breeding Laughing Kookaburra: Feeding Young Is No Laughing Matter," *Animal Behaviour* 59, no. 5 (2000): 1009–1018; S. Legge, "The Effect of Helpers on Reproductive Success in the Laughing Kookaburra," *Journal of Animal Ecology* 69, no. 4 (2000): 714–724.

21. S. Legge and A. Cockburn, "Social and Mating System of Cooperatively Breeding Laughing Kookaburras (*Dacelo novaeguineae*)," *Behavioral Ecology and Sociobiology* 47, no. 4 (2000): 220–229.

22. P. A. Downing, A. S. Griffin, and C. K. Cornwallis, "Hard-working Helpers Contribute to Long Breeder Lifespans in Cooperative Birds," *Philosophical Transactions of the Royal Society B* 376 (2021): 20190742.

Chapter 9: Communal Nesters: Confusion in the Nest

1. C. Riehl, "Evolutionary Origins of Cooperative and Communal Breeding: Lessons from the Crotophagine Cuckoos," *Ethology* 127, no. 10 (2021): 827–836.

2. P. A. Downing, A. S. Griffin, and C. K. Cornwallis, "Group Formation and the Evolutionary Pathway to Complex Sociality in Birds," *Nature Ecology and Evolution* 4, no. 3 (2020): 479–486.

3. J. W. Bradbury and S. L. Vehrencamp, *Principles of Animal Communication* (Sinauer, 1998).

4. S. L. Vehrencamp, "Relative Fecundity and Parental Effort in Communally Nesting Anis, *Crotophaga sulcirostris*," *Science* 197 (1977): 403–405.

5. W. D. Hamilton, "The Genetical Evolution of Social Behaviour," pts. 1 and 2, *Journal of Theoretical Biology* 7, no. 1 (1964): 1–52.

6. R. Dawkins, *The Selfish Gene* (Oxford University Press, 1976).

7. C. Riehl and L. Jara, "Natural History and Reproductive Biology of the Communally Breeding Greater Ani (*Crotophaga major*) at Gatún Lake, Panama," *Wilson Journal of Ornithology* 121, no. 4 (2009): 679–687.

8. C. Riehl and M. J. Strong, "Social Parasitism as an Alternative Reproductive Tactic in a Cooperatively Breeding Cuckoo," *Nature* 567 (2019): 96–99.

9. Riehl, "Evolutionary Origins of Cooperative and Communal Breeding."

10. G. Schmaltz, J. S. Quinn, and C. Lentz, "Competition and Waste in the Communally Breeding Smooth-billed Ani: Effects of Group Size on Egg-laying Behaviour," *Animal Behaviour* 76, no. 1 (2008): 153–162.

11. L. A. Grieves, D. M. Logue, and J. S. Quinn, "Joint-nesting Smooth-billed Anis, *Crotophaga ani*, Use a Functionally Referential Alarm Call System," *Animal Behaviour* 89 (2014): 215–221.

12. J. S. Quinn, R. H. Macedo, and B. N. White, "Genetic Relatedness of Communally Breeding Guira Cuckoos," *Animal Behaviour* 47, no. 3 (1994): 515–529; R. H. Macedo, "Guira Cuckoos: Cooperation, Infanticide, and Female Reproductive Investment in a Joint-nesting Species," in *Cooperative Breeding in Vertebrates: Studies of Ecology, Evolution, and Behavior*, ed. W. D. Koenig and J. L. Dickinson (Cambridge University Press, 2016), 257–272; M. R. Lima et al., "Group Composition, Mating System, and

Relatedness in the Communally Breeding Guira Cuckoo (*Guira guira*) in Central Brazil," *The Auk* 128, no. 3 (2011): 475–486.

13. S.-F. Shen, H.-W. Yuan, and M. Liu, "Taiwan Yuhinas: Unrelated Joint-nesters Cooperate in Unfavorable Environments," in Koenig and Dickinson, *Cooperative Breeding in Vertebrates*, 237–256.

14. S.-F. Shen et al., "Unfavourable Environment Limits Social Conflict in *Yuhina brunneiceps*," *Nature Communications* 3 (2012): 885.

15. C.-C. Lin et al., "A Sequential Collective Action Game and Its Applications to Cooperative Parental Care in a Songbird," *Animal Behaviour* 129 (2017): 151–159.

16. Shen et al., "Unfavourable Environment Limits Social Conflict in *Yuhina brunneiceps*."

17. J. J. Kirchman et al., "Phylogeny Based on Ultra-conserved Elements Clarifies the Evolution of Rails and Allies (Ralloidea) and Is the Basis for a Revised Classification," *The Auk* 138, no. 4 (2021): ukab042.

18. J. L. Craig, "Pair and Group Breeding Behaviour of a Communal Gallinule, the Pukeko, *Porphyrio p. melanotus*," *Animal Behaviour* 28, no. 2 (1980): 593–603.

19. I. G. Jamieson et al., "Shared Paternity Among Non-relatives Is a Result of an Egalitarian Mating System in a Communally Breeding Bird, the Pukeko," *Proceedings of the Royal Society B* 257 (1994): 271–277.

20. C. J. Dey, C. O'Connor, and J. S. Quinn, "Hatching Order Affects Offspring Growth, Survival and Adult Dominance in the Joint-laying Pukeko *Porphyrio melanotus melanotus*," *Ibis* 156, no. 3 (2014): 658–667.

21. J. S. Quinn et al., "Tolerance of Female Co-breeders in Joint-laying Pukeko: The Role of Egg Recognition and Peace Incentives," *Animal Behaviour* 83, no. 4 (2012): 1035–1041.

22. Dey, O'Connor, and Quinn, "Hatching Order Affects Offspring Growth"; Quinn et al., "Tolerance of Female Co-breeders in Joint-laying Pukeko."

23. C. J. Dey, J. Dale, and J. S. Quinn, "Manipulating the Appearance of a Badge of Status Causes Changes in True Badge Expression," *Proceedings of the Royal Society B* 281 (2014): 20132680.

Chapter 10: Supersocial Groups: Birds That Are Always Together

1. J. M. Marzluff and R. P. Balda, *The Pinyon Jay: Behavioral Ecology of a Colonial and Cooperative Corvid* (T. & A. D. Poyser, 1992).

2. G. L. Maclean, "The Sociable Weaver, Part 2: Nest Architecture and Social Organization," *Ostrich* 44 (1973): 191–218.

3. G. L. Maclean, "The Sociable Weaver, Part 1: Description, Distribution, Dispersion and Populations," *Ostrich* 44 (1973): 176–190.

4. Maclean, "The Sociable Weaver, Part 2."

5. Maclean, "The Sociable Weaver, Part 2."

6. G. M. Leighton, "Sex and Individual Differences in Cooperative Nest Construction of Sociable Weavers *Philetairus socius*," *Journal of Ornithology* 155 (2014): 927–935.

7. R. E. van Dijk et al., "The Thermoregulatory Benefits of the Communal Nest of Sociable Weavers *Philetairus socius* Are Spatially Structured Within Nests," *Journal of Avian Biology* 44 (2013): 102–110.

8. G. M. Leighton and S. Echeverri, "Non-linear Influence of Nest Size on Thermal Buffering of Sociable Weaver Nests and the Maintenance of Cooperative Nest Construction," *Avian Biology Research* 7, no. 4 (2014): 255–260.

9. M. Paquet et al., "Communal Roosting, Thermoregulatory Benefits and Breeding Group Size Predictability in Cooperatively Breeding Sociable Weavers," *Journal of Avian Biology* 47 (2016): 749–755.

10. C. N. Spottiswoode, "Phenotypic Sorting in Morphology and Reproductive Investment Among Sociable Weaver Colonies," *Oecologia* 154, no. 3 (2007): 589–600.

11. G. M. Leighton et al., "Relatedness Predicts Multiple Measures of Investment in Cooperative Nest Construction in Sociable Weavers," *Behavioral Ecology and Sociobiology* 69, no. 11 (2015): 1835–1843.

12. R. Covas et al., "Kin Associations and Direct vs Indirect Fitness Benefits in Colonial Cooperatively Breeding Sociable Weavers *Philetairus socius*," *Behavioral Ecology and Sociobiology* 60, no. 3 (2006): 323–331.

13. P. B. D'Amelio et al., "Benefits of Pair-bond Duration on Reproduction in a Lifelong Monogamous Cooperative Passerine," *American Naturalist* 203, no. 5 (2024): 576–589.

14. R. Covas, M. A. du Plessis, and C. Doutrelant, "Helpers in Colonial Cooperatively Breeding Sociable Weavers *Philetairus socius* Contribute to Buffer the Effects of Adverse Breeding Conditions," *Behavioral Ecology and Sociobiology* 63, no. 1 (2008): 103–112.

15. P. B. D'Amelio et al., "Disentangling Climatic and Nest Predator Impact on Reproductive Output Reveals Adverse High-temperature Effects Regardless of Helper Number in an Arid-region Cooperative Bird," *Ecology Letters* 25, no. 1 (2022): 151–162.

16. R. Covas, C. Doutrelant, and M. A. du Plessis, "Experimental Evidence of a Link Between Breeding Conditions and the Decision to Breed or to Help in a Colonial Cooperative Bird," *Proceedings of the Royal Society B* 271 (2004): 827–832.

17. R. Covas and M. A. du Plessis, "The Effect of Helpers on Artificially Increased Brood Size in Sociable Weavers (*Philetairus socius*)," *Behavioral Ecology and Sociobiology* 57, no. 6 (2005): 631–636.

18. A. M. Lowney and R. L. Thomson, "Ecological Engineering Across a Temporal Gradient: Sociable Weaver Colonies Create Year-round Animal Biodiversity Hotspots," *Journal of Animal Ecology* 90 (2021): 2362–2376; A. M. Lowney and R. L. Thomson, "Ecological Engineering Across a Spatial Gradient: Sociable Weaver Colonies Facilitate Animal Associations with Increasing Environmental Harshness," *Journal of Animal Ecology* 91 (2022): 1385–1399.

19. T. K. Aikins, M. D. Cramer, and R. L. Thomson, "Positive Feedbacks Between Savanna Tree Size and the Nutritional Characteristics of 'Islands of Fertility' Are Amplified by Sociable Weaver Colonies," *Journal of Arid Environments* 209 (2023): 104903.

20. Lowney and Thomson, "Ecological Engineering Across a Temporal Gradient"; Lowney and Thomson, "Ecological Engineering Across a Spatial Gradient."

21. A. M. Lowney and R. L. Thomson, "Costs and Benefits in Extreme Nesting Associations: Do Sociable Weavers Benefit from Hosting African Pygmy Falcons?," *Ibis* 166, no. 3 (2024): 801–813.

22. T. L. Rymer, R. L. Thomson, and M. J. Whiting, "At Home with the Birds: Kalahari Tree Skinks Associate with Sociable Weaver Nests Despite African Pygmy Falcon Presence," *Austral Ecology* 39, no. 7 (2014): 839–847.

23. W. D. Koenig et al., "Acorn Woodpecker (*Melanerpes formicivorus*)," version 1.0, in *Birds of the World*, ed. P. G. Rodewald and B. K. Keeney (Cornell Lab of Ornithology, 2020).

24. W. D. Koenig et al., "Acorn Woodpecker (*Melanerpes formicivorus*)," in Rodewald and Keeney, *Birds of the World*.

25. W. D. Koenig et al., "Patterns and Consequences of Egg Destruction Among Joint-nesting Acorn Woodpeckers," *Animal Behaviour* 50, no. 3 (1995): 607–621.

26. W. D. Koenig, "The Incidence of Runt Eggs in Woodpeckers," *Wilson Bulletin* 92, no. 2 (1980): 169–176.

27. W. D. Koenig et al., "Are You My Baby? Testing Whether Paternity Affects Behavior of Cobreeder Male Acorn Woodpeckers," *Behavioral Ecology* 32, no. 5 (2021): 865–874.

28. W. D. Koenig, *Population Ecology of the Cooperatively Breeding Acorn Woodpecker* (Princeton University Press, 1987).

29. W. D. Koenig and J. Haydock, "Oaks, Acorns, and the Geographical Ecology of Acorn Woodpeckers," *Journal of Biogeography* 26, no. 1 (1999): 159–165.

30. P. B. Stacey and C. E. Bock, "Social Plasticity in the Acorn Woodpecker," *Science* 202 (1978): 1298–1300; J. D. Ligon and P. B. Stacey, "Land Use, Lag Times and the Detection of Demographic Change: The Case of the Acorn Woodpecker," *Conservation Biology* 10, no. 3 (1996): 840–846.

31. Ligon and Stacey, "Land Use, Lag Times and the Detection of Demographic Change."

32. R. G. Heinsohn, A. Cockburn, and R. B. Cunningham, "Foraging, Delayed Maturation, and Advantages of Cooperative Breeding in White-winged Choughs, *Corcorax melanorhamphos*," *Ethology* 77, no. 3 (1988): 177–186.

33. K. A. Jønsson et al., "A Supermatrix Phylogeny of Corvoid Passerine Birds (Aves: Corvides)," *Molecular Phylogenetics and Evolution* 94 (2016): 87–94.

34. Heinsohn, Cockburn, and Cunningham, "Foraging, Delayed Maturation, and Advantages of Cooperative Breeding"; R. G. Heinsohn et al., "Coalitions of Relatives and Reproductive Skew in Cooperatively Breeding White-winged Choughs," *Proceedings of the Royal Society B* 267 (2000): 243–249.

35. R. G. Heinsohn, "Inter-group Ovicide and Nest Destruction in Cooperatively Breeding White-winged Choughs," *Animal Behaviour* 36, no. 6 (1988): 1856–1858.

36. R. G. Heinsohn and A. Cockburn, "Helping Is Costly to Young Birds in Cooperatively Breeding White-winged Choughs," *Proceedings of the Royal Society B* 256 (1994): 293–298.

37. R. G. Heinsohn, "Kidnapping and Reciprocity in Cooperatively Breeding White-winged Choughs," *Animal Behaviour* 41, no. 6 (1991): 1097–1100.

38. C. R. J. Boland, R. G. Heinsohn, and A. Cockburn, "Experimental Manipulation of Brood Reduction and Parental Care in Cooperatively Breeding White-winged Choughs," *Journal of Animal Ecology* 66, no. 5 (1997): 683–691.

39. C. R. J. Boland, R. G. Heinsohn, and A. Cockburn, "Deception by Helpers in Cooperatively Breeding White-winged Choughs and Its Experimental Manipulation," *Behavioral Ecology and Sociobiology* 41, no. 4 (1997): 251–256.

40. K. Johnson and R. P. Balda, "Pinyon Jay (*Gymnorhinus cyanocephalus*)," version 2.0, in *Birds of the World*, ed. P. G. Rodewald and B. K. Keeney (Cornell Lab of Ornithology, 2020); K. Johnson and G. Sadoti, "A Review of Pinyon Jay (*Gymnorhinus cyanocephalus*) Habitat Ecology," *Wilson Journal of Ornithology* 135, no. 2 (2023): 232–247.

41. Johnson and Balda, "Pinyon Jay (*Gymnorhinus cyanocephalus*)," in Rodewald and Keeney, *Birds of the World*.

42. K. Johnson et al., "Home Range- and Colony-scale Habitat Models for Pinyon Jays in Piñon-Juniper Woodlands of New Mexico, USA," *Avian Conservation and Ecology* 11, no. 2 (2016): 6.

43. Johnson and Sadoti, "A Review of Pinyon Jay (*Gymnorhinus cyanocephalus*) Habitat Ecology."

44. J. M. Marzluff and R. P. Balda, *The Pinyon Jay: Behavioral Ecology of a Colonial and Cooperative Corvid* (T. & A. D. Poyser, 1992).

45. Marzluff and Balda, *The Pinyon Jay: Behavioral Ecology of a Colonial and Cooperative Corvid.*

46. K. Johnson, "Sexual Selection in Pinyon Jays I: Female Choice and Male–Male Competition," *Animal Behaviour* 36, no. 4 (1988): 1038–1047; K. Johnson, "Sexual Selection in Pinyon Jays II: Male Choice and Female–Female Competition," *Animal Behaviour* 36, no. 4 (1988): 1048–1053.

47. Johnson and Balda, "Pinyon Jay (*Gymnorhinus cyanocephalus*)," in Rodewald and Keeney, *Birds of the World.*

48. J. M. Marzluff, "Do Pinyon Jays Alter Nest Placement Based on Prior Experience?," *Animal Behaviour* 36, no. 1 (1988): 1–10.

49. R. P. Balda and A. C. Kamil, "A Comparative Study of Cache Recovery by Three Corvid Species," *Animal Behaviour* 38, no. 3 (1989): 486–495.

50. J. D. Boone, E. Ammon, and K. Johnson, "Long-term Declines in the Pinyon Jay and Management Implications for Piñon-Juniper Woodlands," in *Trends and Traditions: Avifaunal Change in Western North America*, ed. W. D. Shuford, R. E. Gill Jr., and C. M. Handel (Western Field Ornithologists, 2018), 190–197.

51. R. M. Lanner, *The Piñon Pine: A Natural and Cultural History* (University of Nevada Press, 1981).

52. W. H. Romme et al., "Historical and Modern Disturbance Regimes, Stand Structures, and Landscape Dynamics in Pinon-Juniper Vegetation of the Western U.S.," *Rangeland Ecology and Management* 62, no. 3 (2009): 203–222.

53. Boone, Ammon, and Johnson, "Long-term Declines in the Pinyon Jay."

54. P. A. Magee, J. D. Coop, and J. S. Ivan, "Thinning Alters Avian Occupancy in Piñon-Juniper Woodlands," *The Condor* 121, no. 1 (2019): duy008.

55. A. P. Wion et al., "Aridity Drives Spatiotemporal Patterns of Masting Across the Latitudinal Range of a Dryland Conifer," *Ecography* 43, no. 4 (2020): 569–580.

56. M. D. Redmond, F. Forcella, and N. N. Barger, "Declines in Pinyon Pine Cone Production Associated with Regional Warming," *Ecosphere* 3, no. 12 (2012): 1–14.

57. D. D. Breshears et al., "Regional Vegetation Die-off in Response to Global-change-type Drought," *Proceedings of the National Academy of Sciences* 102, no. 42 (2005): 15144–15148.

Conclusion: Why Are Birds So Social?

1. K. V. Rosenberg et al., "Decline of the North American Avifauna," *Science* 366 (2019): 120–124.

Index

A

Acacia Pied Barbets, 218
Acorn Woodpeckers, 210, 218–27, 235–36, 244–45
Adams, Amy Yackel, 43
adaptive traits, 91
Adélie Penguins, 109, 110, 113
African Harrier Hawks, 33
African Hobbies, 33, 34, 35
African Pygmy Falcons, 218
African Wood Owls, 33
Agashe, Deepa, 56, 185
Ainley, David, 109–10
alarm calls and papá-uirá, 48–51, 54–55, 57–62, 67, 68, 247–49
albatrosses, 98, 110
Alejandro, Juan, 155
Alexander, Richard D., 72–73, 75, 78
Allen, Arthur, 27–28
Allen, Robert, 93, 94
alpha–beta male dynamic, 131–36, 248
altruistic behaviors, 5, 21
American Coots, 8
American Kestrels, 59, 84
American Robins, 87–88
Anderson, David, 101–2
anis, 192–200, 207
Anthony, Robert, 42
antshrikes, 46, 50, 51, 61, 62, 64, 67
antwrens, 46, 50, 51, 61, 62, 67
Aplomado Falcons, 60
Archaeopteryx fossil, 141
Ashmole, Philip, 104–5
Ashmole's halo, 105, 110
Ashy Drongos, 54, 55
Ashy Tits, 218
Atlantic Puffins, 3, 102–4, 105, 107
Attenborough, David, 56
Attwater Prairie-Chicken, 116–17
auks, 98, 112
Australian Brushturkeys, 145

B

Back-legged Kittiwakes, 103, 104
Balda, Russell, 236–37, 238, 239, 240, 241
Bald Eagles, 42–43, 247
Bank Swallows, 70, 74–76, 77, 80, 87
Barn Owls, 33, 84
Barn Swallows, 32–33, 34–35, 247
Bar-tailed Godwits, 10
Bartlett, Katharine, 237
Bateman, Gary, 236–37
Bates, Henry Walter, 47–49, 68
Beauchamp, Guy, 15–17, 44
Belted Kingfishers, 77
Bensch, Staffan, 153
Bicolored Hawks, 60
Bijlsma, Rob, 32–35
Biological Dynamics of Forest Fragments Project, 66
birds-of-paradise, 228
Birds of the World, 43
bird-trade business, 65–66
Birkhead, Tim, 102, 105
Black-and-white Warblers, 53
Black-billed Magpies, 84
blackbirds, 8, 14, 16, 18, 26, 27–29, 72, 90, 146–47
Black-capped Chickadees, 46, 52, 53, 59–60

Index

Black Guillemots, 103, 104
Black-headed Cuckooshrikes, 55
Black-legged Kittiwakes, 105, 107
Black-naped Monarchs, 54
Black Noddy birds, 104
Black-throated Green Warblers, 53
Black Tits, 51
Black Vultures, 37–42, 247
bluebirds, 167
Blue-footed Booby, 101–2
Blue Jays, 77
Blue Tanagers, 49
Blue Tits, 159–60
Bluish-slate Antshrikes, 46, 51, 61
Boag, Peter, 147
Bock, Carl, 225
Boland, Christopher, 230, 231–32
boobies, 98, 101–2
Borgia, Gerry, 118
Botero, Carlos, 183
Bradbury, Jack, 122–23, 194
Bray, Olin, 146–47
Brazilian National Institute of Amazonian Research, 66
breeding. *See* mating and reproducing
Breininger, David, 175–76
Brewer's Blackbirds, 90
Briskie, James, 167
Broad-winged Hawks, 22–24
Bronzed Drongos, 54
Bronze Manakins, 31–32
Brouwer, Lyanne, 148
Brown, Charles, 79–87, 93
Brown, Mary Bomberger, 79–87
Brown-bellied Stipple-throats, 62
Brown-cheeked Fulvetta, 55, 56
Brown-headed Cowbirds, 29
Brown-headed Nuthatches, 53
Brown Pelicans, 8, 21
brushturkeys, 145
Buitron, Deborah, 91
buntings, 149–50, 167, 248
Burger, Joanna, 72
Burley, Nancy, 158–59
Bushtits, 27, 29–30

C

Calhim, Sara, 157
cardinals, 140
Carolina Chickadees, 52
Carter, Geoffrey, 175–76
Cassin's Auklets, 113
cassowaries, 143
Catchpole, Clive, 151, 152
cattle, 13–14, 116
Cattle Egrets, 71, 72
Cedar Waxwings, 8, 36
Chaplin, Susan, 30
Checker-throated Antwrens, 50
Chestnut-rumped Woodcreepers, 62, 67
Chiappe, Luis, 141
chickadees, 46, 52, 53, 59–60, 247
chicks
  advantages of helpers for, 174, 181, 186, 188–90, 217
  begging of, 166–67
  competition of, 164, 188
  and early-vs.-last laid eggs, 195–96
  fledging of, 167
  kidnapped by competing neighbors, 229–30
  parasite, 80–82, 96, 167
  predation on, 173–74, 214
  survival rates of, 181, 203, 226
  See also parental care
Chinstrap Penguins, 109, 111
Christmas Bird Counts, 223–24
Cinereous Antshrikes, 46, 62, 67
Cliff Swallows, 3, 70, 79–87
climate change, 111, 112–13, 244
Coal Tits, 57
Cockburn, Andrew, 146, 156, 176, 179, 187
colonies, 69–96
  advantages of, 3, 77, 89, 96, 101
  costs/disadvantages of, 73–76, 80–82, 95–96, 214
  density of nests in, 72
  egg parasitism, 80–82, 96
  egg tossing/dumping behaviors in, 80–82
  and evolutionary changes, 85–87
  and foraging, 76, 82–83, 90–91, 96, 101
  parental care in, 3–4
  and predators, 72, 77–78, 83–85, 90, 95, 101–4, 214
  quality of habitat as motive for, 91–93
  of seabirds, 101–4
  as successful nesting strategy, 70
Common Crows, 77
Common Cuckoos, 152
Common Gallinules, 162, 204
Common Grackles, 14, 26, 84, 170

Common Guillemots (Murres), 102, 103, 104, 107, 112
Common Redshank, 9–14, 29, 247
communal breeding, 191–207
advantages of, 194–95
of Greater Anis, 192, 196–98, 200
of Groove-billed Anis, 192, 193, 194, 195–96
of Guira Cuckoos, 193, 196, 199, 200–201, 207
of Pukekos, 204–7
of Taiwan Yuhinas, 201–4
communal roosts, 2, 25–44
advantages of, 29–31
arrivals and departures from, 33–34
of Bald Eagles, 42–43
disadvantages of, 27
dominance hierarchies in, 31–32
and foraging, 35–42, 44
multiple species in, 28–29
and nesting, 26
and polyandry, 193
positions of birds in, 28–29, 31–32
and predators, 28, 32–35
sizes of, 26–27, 28–29
and sleeping of birds, 26
competition among birds, 140, 188, 203–4, 205, 228, 229–30
conflict, zone of, 166
contact calls, 60
cooperative behaviors, 5, 182–83, 194–95, 249
cooperative breeding, 172
Cooper's Hawks, 59
coots, 8, 204
cormorants, 71, 98, 105, 111
Cornwallis, Charlie, 181, 183, 189
Coulson, John, 104
courtship of Pinyon Jays, 237–38
Covas, Rita, 215
cowbirds, 29, 167
Craig, John, 205
Cresswell, Will, 10–11, 12
Crested Tits, 56, 57
crotophagine cuckoos, 192–93
crows, 77, 89, 180
cuckoos, 152, 192–93, 196, 199, 200–201, 207

D

Damaraland White-eye, 51
D'Amelio, Pietro, 215, 216
Dark-fronted Babblers, 55
Darwin, Charles, 141, 161
Davidson, Amy, 141–42
Dawkins, Richard, 73, 165, 194
Dey, Cody, 206–7
die-offs of birds (wrecks), 112–13
"differential-allocation hypothesis," 158, 159
dinosaurs, 141, 145
diving petrels, 98
Dolby, Andrew, 58
dominance hierarchies, 31–32, 202–3, 205, 206–7, 231
Dot-winged Antwrens, 50
Double-crested Cormorants, 105
Doutrelant, Claire, 215
doves, 77
Downing, Philip, 189, 193
Downy Woodpeckers, 46, 52, 53, 59, 60, 248
drongos, 54, 55
Dunnocks, 167
Dusky-throated Antshrikes, 51, 61, 62
DuVal, Emily, 116, 130–37

E

Eastern Bluebirds, 167
Eastern Kingbirds, 93
eBird, 27–28
ectoparasites, 27, 75, 81, 82, 85–86, 87
eggs
of Acorn Woodpeckers, 221, 222–23
in communal nests, 204–5
destruction of, 222–23
early-vs.-last laid, 195–96, 205
egg parasitism, 80–82, 96, 198
egg-tossing/dumping behaviors, 80–82, 198, 200–201
helpers and size of, 179–80
paternal behavior and color of, 162–63
runt, 222
egrets, 37, 71, 72
El Niño years, 107
Emperor Penguins, 109, 113
emus, 143
endangered species, 64
environmental damage, 63–68
Eurasian Coots, 204
Eurasian Golden Orioles, 55
Eurasian Marsh Harriers, 33
Eurasian scops-owls, 33
Eurasian Sparrowhawks, 10–12
Eurasian Treecreepers, 57

Index

European Shags (cormorants), 105
European Starlings, 14, 26, 29, 32
evolutionary changes, 5, 86–87, 118, 141
Ezenwa, Vanessa, 131

F

fairywrens, 156–57, 176–80
falcons, 33, 60, 89, 218
Farine, Damien, 65
Fernández, Gustavo, 144
Fieldfares, 87–90, 248
finches, 15, 110, 159, 218
fires, 174–76
Fitzpatrick, John, 170–71, 172
fledging of young, 167–68
flightless birds, 143
flocks, 7–24
  advantages of, 2, 8–9, 11
  characteristics of, 15–17
  disadvantages of, 8, 12–13
  in flight, 17–19
  foraging of, 9–11, 12–13, 15–16
  migrations in, 17, 19–24
  names for, 15
  and predators, 8–9, 11–13, 16–17
  See also mixed-species flocks
Florida Scrub-jays, 172–76
flycatchers, 51, 54, 55
flying in flocks, 17–19
food and foraging, 10–11
  and Acorn Woodpeckers, 223–25
  and alarm calls of birds, 61–62
  of birds in colonies, 76, 82–83, 90–91, 96, 101
  and communal roosts, 35–42, 44
  depletion of (Ashmole's halo), 105, 110
  finding alternate sources of, 236
  of flocks, 9–11, 12–13, 15–16
  helpers' assistance with, 188–89, 216–17, 230, 231
  helpers' consumption of, 232
  helpers related to availability of, 225–26
  and "information center hypothesis," 36–37, 39, 41, 43
food and foraging (*cont.*)
  and "local enhancement hypothesis," 35–36, 41
  of mixed-species flocks, 53–55, 57, 59, 61–62, 64, 67
  and nesting sites, 104–5
  and predators, 54–55
  and sallyers, 54–56
  of seabirds, 104–8, 111
  and strategies for feeding chicks, 231
  synchronous feeding, 203
  times of scarcity, 213, 223, 224–25, 231, 235–36
Foster, Gene, 237
frigate birds, 98
fulmars, 98
Furness, Robert, 105, 106

G

Galápagos Hawks, 101–4
Galápagos Islands, 16, 18, 98, 101–4, 106
gannets, 98
Gauthreaux, Sidney, Jr., 23–24
genes and chromosomal inversions, 127–29
Gentoo Penguins, 109, 110
"Geometry for the Selfish Herd" (Hamilton), 8, 11, 19
Gibson, Rob, 122–23
Goldcrests, 56–57
Golden-crowned Kinglets, 53–54
Golden Eagles, 121
Golden-masked Tanagers, 49
Gomez, Juan, 63
grackles, 14, 26, 84, 146, 153–56, 170
Gradwohl, Judy, 50
Gray Antwrens, 46, 51, 62
Gray Catbirds, 163
Great Black-backed Gulls, 103
Great Blue Herons, 37
Great Egrets, 37, 71
Greater Anis, 4, 192, 196–98, 200
Greater Racket-tailed Drongo, 54
Greater Rheas, 143–44
Greater Sage-grouse, 3, 118–24
Great Horned Owls, 84
Great Kiskadees, 23
Great Reed Warblers, 151–52, 153
Great-tailed Grackles, 146, 153–56
Great Tits, 57
Great White Pelicans, 20
grebes, 91–92
Greenberg, Russell, 50
Green-capped Eremomela, 51
Green Honeycreepers, 49
Green Jays, 23
Green Kingfishers, 23
Green Woodhoopoes, 189
Grey-cheeked Bulbul, 65
Grey-cheeked Fulvettas, 46
Griesser, Michael, 183, 184
Griffin, Ashleigh, 181–82, 184, 185, 189

Griffith, Simon, 148
Groove-billed Anis, 192, 193, 194, 195–96
Grubb, Thomas, Jr., 58
guillemots, 98, 102
Guira Cuckoos, 193, 196, 199, 200–201, 207
gulls, 98, 101, 103–4

H

Haas, Volker, 88
habitat loss, 62
Hairy Woodpeckers, 59
Hamilton, Bill, 8–9, 11, 12, 19, 164–65, 194
Hamilton-Zuk hypothesis, 118
Hanley, Daniel, 162–63
Harris's Hawks, 60
Hasselquist, Dennis, 153
Hatchwell, Ben, 30–31, 180
hawks, 10–13, 22–24, 33, 55, 59, 60, 77, 101–4, 247
Haydock, Joseph, 223–24
Heinsohn, Robert, 227, 228, 229–30, 231–32
helpers, 169–90
  and Acorn Woodpeckers, 221
  advantages afforded by, 171–72, 174, 181, 184, 186, 188–90, 216, 240
  and cooperative living among birds, 182–83
  dependence on, 228–30, 232
  and egg size, 179–80
  and food availability, 188–89, 216–17, 225–26, 230, 231
  food secretly consumed by, 232
  impulses driving, 133–34, 170–71
  and kin selection, 182
  and monogamous mating, 193
  and number of fledglings, 186
  origins of, 183–84
  and Pinyon Jays, 239–40
  and predators, 173–74, 190
  primary/secondary, 185–86
  of Sociable Weavers, 215–16
  transition to breeders, 178–79, 230, 240
Herbert-Read, James, 19
herons, 37, 71
Herre, Allen, 192
Herring Gulls, 103
High Island, 70–72
Hodum, Peter, 101–2
Hogan-Warburg, Alida, 125–27
homing pigeons, 17–19
Hooded Crows, 89
Hoogland, John, 74–76, 77, 78–79, 80, 82
Horn, Henry, 90–91
House Finches, 15
House Sparrows, 140
Hoysak, Drew, 28
Hughes, Colin, 91
human impact on wildlife, 63–68

I

Indian Paradise-flycatchers, 54
Indigo Buntings, 149–50, 167, 248
"information center hypothesis," 36–37, 39, 41, 43

J

jaegers, 98
jays, 4, 5, 23, 77, 172–76, 180, 210, 232–33, 237–38
Johnsen, Arild, 160, 161
Johnson, Kris, 131, 154, 155, 210, 232–33, 234, 235, 236, 238, 242
Jones, Todd, 167
Jukema, Joop, 128–29
Jullien, Mathilde, 62, 63

K

Kacelnik, Alex, 15
Kamil, Alan, 241
Karoo Scrub-robins, 189
Keister, George, Jr., 42
Kerlinger, Paul, 23–24
kestrels, 59, 77, 84
kingbirds, 93
kingfishers, 23, 70, 77, 184–86
kin selection, 5, 164–65, 170–71, 182, 194
kiwis, 143
Koenig, Walt, 172, 219, 220, 222, 223–24, 225
kookaburras, 185, 187–88
Krakauer, Alan, 123–24, 137
Kunte, Krushnamegh, 56, 185

L

Lack, David, 101
Lamey, Tim, 36
Lance-tailed Manakins, 3, 130–36, 248
Lank, David "Dov," 124–25, 127, 130, 137
Lanner, Ronald, 242–43
Lanner Falcons, 33

Lapland Longspurs, 167
Laughing Kookaburra, 185, 187–88
Leach's Storm-petrels, 103, 104
Leech, Susan, 167
Legge, Sarah, 187–88
Leighton, Gavin, 212–13, 214
leks and lekking species, 3, 115–37
  about leks, 116–18
  contributors to mating success, 118–24
  function of, 117–18
  and multiple matings/paternity, 130–31
  paradox of, 117–18
  and predators, 121
  and sex of offspring, 136–37
  See also mating and reproducing
Lesser Flamingos, 15
Ligon, David, 225, 226
Little Blue Herons, 71
"local enhancement hypothesis," 35–36, 41
logging operations, 63–64
Long-billed Antwrens, 62
longspurs, 167
Long-tailed Tits, 5, 30–31, 180, 248
Long-winged Antwrens, 51
Lovejoy, Thomas, 66
Lowney, Anthony, 218

M

Macedo, Regina, 200
Maclean, Gordon Lindsay, 211
Magee, Patrick, 243
Magnolia Warblers, 53
magpies, 84
Malpas, Lucy, 13
Manx Shearwaters, 99–101
Marthy, William, 65
Martínez, Ari, 60, 61, 63
Marzluff, John, 238, 241
mating and reproducing
  of Acorn Woodpeckers, 221
  alpha–beta male dynamic in, 131–36, 248
  breeding rights, 4
  communal breeding, 4, 201–2
  competition for, 73, 203–4, 205, 223
  contributors to success in, 118–24
  cooperative breeding, 172
  courtship of Pinyon Jays, 237–39
  and degree of relatedness, 222–23
  dominance hierarchies in, 205, 206–7
  genetic impacts on success of, 127–29
  of helpers-turned-breeders, 178–79
  increasing odds of, 5
  and kin selection, 164–65, 183, 194
  male quality and female choice, 118, 123, 133, 151, 153, 156, 157, 158–61
  mate attractiveness, 136–37
  monogamy, 143, 148, 163, 183, 193, 215
  multiple matings/paternity, 120, 130–31, 133, 146–47
  and nest sites, 103–4
  outside pair bond, 4, 146–51, 152–53, 155–57, 167, 168, 178, 248
  polyandry, 146, 193
  polygyny, 146, 147, 163–64
  of Pukekos, 205
  rate of success for rheas, 144
  of seabirds, 110–11
  and sex of offspring, 136–37, 160
  and skew theory, 193
  of songbirds, 110–11
  songs of males, 151–53
  See also leks and lekking species; parental care
McGowan, Andrew, 31
Meanley, Brooke, 28–29
megapodes, 145
Melampittidae, 228
Merlins, 27–28, 89
migrations, 17, 19–24, 70–71
mixed-species flocks, 45–68
  advantages of, 2–3
  and contact calls, 60
  endangered species, 64
  foraging of, 53–55, 57, 59, 61–62, 64, 67
  fragility of, 67–68
  human impact on, 63–68
  and papá-uirá/alarm calls of birds, 48–51, 54–55, 57–62, 67, 68, 247–48
  and predators, 50–51, 53–54, 56–62
  and sallyers, 54–56
Mock, Doug, 36, 37
Mokross, Karl, 66–67
monogamy, 143, 148, 163, 183, 193, 215
Morse, Douglass, 52–53
Morton, Eugene, 95
Mourning Doves, 77
Moynihan, Martin, 49–50
Mumme, Ron, 172–73, 223
Munn, Charles, 51
murres, 112
Musters, Robert, 18
Myrtle Warblers, 52

N

names for flocks of birds, 15
Native Americans, 48, 93
natural selection, 141, 161–62, 170–71, 194
Naugler, Christopher, 167
Neotropic Cormorants, 71
nests and nesting
  of Common Redshank, 13–14
  costs of incubation, 229
  destruction of competitors' nests, 228
  failure rate of, 216
  and food availability, 104–5
  and population sizes, 104–5
  and predators, 72, 77–78, 83–85, 90, 95, 101, 102–4, 173–74, 180, 198, 199–200, 216
  and roosts, 26
  of seabirds, 101–6
  site choices for, 104–6
  of Sociable Weavers, 211–12
  See also colonies; communal breeding; eggs
Nettleship, David, 102–3
Nice, Margaret Morse, 93, 94
Northern Bald Ibis, 20
Northern Cardinals, 140
Northern Gannets, 105
Noske, Richard, 32
nuclear birds, 50, 67
Nuechterlein, Gary, 91
nuthatches, 46, 52, 53, 58, 59, 248

O

ocean conditions, 106–8
Orange-footed Megapodes, 145
Orange Minivets, 55, 56
ostriches, 143
Oviraptor fossil, 141, 143
owls, 33, 57, 84

P

Palm Tanagers, 49
papá-uirá/alarm calls of birds, 48–51, 54–55, 57–62, 67, 68
Paquet, Matthieu, 213
Paradise-flycatchers, 55
parasite eggs/chicks, 80–82, 96, 167, 198
parental care, 3–4
  and begging of chicks, 166–67
  biparental, 146
  competition of chicks for, 164
  female-only, 146
  investment of parents in brood, 157–65
  male-only, 141–46, 193
  and parasite eggs/chicks, 80–82, 96, 167
  peacocks' abstinence from, 161
  and predators, 155
  stronger young favored in, 166
  See also helpers
Parker, Patricia (Patty), 38–42
Patricelli, Gail, 120, 123, 137
peacocks, 3, 161–62
pelicans, 8, 20, 21, 98, 111
penguins, 5, 106, 109–10, 111, 113
Peters, Anne, 179
petrels, 98, 103, 104, 110, 111, 112
Petrie, Marion, 162
Piatt, John, 112
Pied Kingfishers, 70, 184–86
Piersma, Theunis, 128–29
pigeons, 17–19
Pine Warblers, 53
Pinyon Jays, 4, 5, 210, 232–33
pinyon pine trees, 233–34, 235, 242–43, 244
Plain-colored Tanagers, 49–50
Plovers, 10
polyandry, 146, 193
polygyny, 146, 147, 163–64
population declines, 249
predators
  and advantages of flocking, 8–9, 11–13, 16–19
  and alarm calls of birds, 50–51, 59–60, 61–62, 67, 68, 247–49
  and begging of chicks, 167
  and colonies, 214
  and communal roosting, 28, 32–35
  Fieldfares's responses to, 88
  and fledging of young, 167–68
  and foraging of birds, 54–55
  helpers' alerts about, 173–74, 190
  islands lacking, 16–19
  and lekking grounds, 121
  and male grackles, 155
  and mixed-species flocks, 50–51, 53–54, 56–62
  nests attacked by, 72, 77–78, 83–85, 90, 95, 101, 102–4, 173–74, 180, 198, 199–200, 216
  and papá-uirá/alarm calls of birds, 48–51, 54–55, 57–62, 67, 68

predators (*cont.*)
  and seabird colonies, 101–4, 112
  snakes, 84–85, 90
puffins, 3, 98, 102–4, 105, 107
Pukekos, 204–7
Purple Martins, 93–96
Pygmy Owls, 57

Q

quelea, 27, 70
Quinn, John, 11–12, 199

R

rats, 112
Razorbills, 103
Reboreda, Juan, 144
Red-billed Quelea, 70
Red-breasted Nuthatches, 53
Red-breasted sparrowhawks, 33
Red-cockaded Woodpecker, 189
Red-necked Grebes, 91–92
Red-tailed Tropicbirds, 111
Red-winged Blackbirds, 8, 14, 16, 18, 26, 27–29, 72, 146–47
Reyer, Heinz-Ulrich, 185–86
rheas, 143–44
Richards, Del, 145
Riehl, Christie, 196–98
roadrunners, 193
robins, 87–88, 189
robot birds, 18–19, 123–24, 137
Rock Pigeons, 17
Rooks, 32
Roseate Spoonbills, 71
Rose-breasted Grosbeaks, 71
Ruby-crowned Kinglets, 52
Ruffs, 3, 124–30, 248
Rufous-rumped Foliage-gleaners, 62
Rufous-tailed Foliage-gleaners, 51
Russell, Andy, 179–80
Rusty Blackbirds, 26, 29
Rymer, Tasmin, 218

S

Sachs, Joel, 91, 92
sallyers, 54–56, 67
sandpipers, 9–10
Sandvik, Hanno, 107–8
Sankey, Daniel, 17
Sansom, Alex, 12
Sardell, Rebecca, 137
Scaly-feathered Finches, 218
Scarlet Minivets, 55, 56
Scarlet Tanagers, 71
Scott, John, 119–21
scrub-jays, 172–76
seabirds and seabird colonies, 3, 97–113
  breeding, 110–11
  and climate change, 112–13
  declines in, 111
  food availability for, 104–8, 111, 248
  four orders of seabirds, 98
  humans' impacts on, 111–12
  and Manx Shearwaters, 99–101
  nesting patterns of, 101–4
  and ocean conditions, 106–8
  predation on, 101–4, 112
  size of colonies, 104–5
The Selfish Gene (Dawkins), 73, 165, 194
Semple, Katy, 124
Senegal Coucals, 33
Sharps, Elwyn, 13–14
Sharp-shinned Hawks, 58, 59
shearwaters, 98, 99–101, 110, 111
Sheldon, Ben, 159
Shen, Sheng-Feng, 201–4
Shepard, Emily, 22–24
Sherman, Paul, 74–76, 77, 80, 82
Shikra, 33
shorebirds, 14
skuas, 98
Slaty Antshrikes, 50
sleep, 26. *See also* communal roosts
Smith, Connie, 130
Smooth-billed Anis, 192, 199
snakes, 84–85, 90, 198, 214, 216
Snow Geese, 21
Snowy Egrets, 71
Sociable Weavers, 210–18, 244–45
Soler, Juan, 163, 164
songbirds, 23, 53, 64, 65–66, 70–71, 110–11, 164, 167, 248. *See also specific species*
songs of males, 151–53, 202–3
Song Sparrows, 77
Sooty Terns, 104
Southern Helmet-shrikes, 51
sparrowhawks, 11–13, 77, 247
sparrows, 59, 77, 140, 167
spoonbills, 71
Spottiswoode, Claire, 213–14
Sridhar, Hari, 54, 55, 56
Stacey, Peter, 225, 226
starlings, 14, 26, 29, 32
Steadman, David, 111, 112

storm-petrels, 98, 112
Stutchbury, Bridget, 93–95
Suhonen, Jukka, 56, 57
Sullivan, Kimberly, 58–59
Superb Fairywrens, 156–57, 176–80
supersocial birds, 4–5, 209–45
Acorn Woodpeckers, 218–27, 244–45
Pinyon Jays, 232–33
as related to resources, 225–27
and smaller vs. larger colonies, 213–14
Sociable Weavers, 210–18, 244–45
White-winged Chough, 227–32, 244–45
swallow bugs, 81–82, 85–86, 87
swallows, 3, 32–33, 34–35, 70, 74–76, 77, 79–87, 167, 247

T

Taiwan Yuhinas, 201–4, 207
tanagers, 49–50, 71
Terborgh, John, 51
terns, 98, 104
Thiollay, Jean-Marc, 62, 63–64
Thompson, Desmond, 36
Thomson, Robert, 218
thrushes, 110
tinamous, 143
titmice, 46, 52, 54, 58, 59, 60, 247
tits, 5, 30–31, 56, 57, 159–60, 180, 218, 248
Torres-Sovero, Claudio, 46
traded bird species, 65–66
Tree Swallows, 167
Tricolored Herons, 71
Trivers, Robert, 165–66, 182
Troodon fossil, 143
Tufted Titmice, 46, 52, 54, 58, 59, 60, 247
Turkey Vultures, 37

U

Usherwood, James, 19

V

Van den Brink, Bennie, 32–35
Van Dijk, René, 212
Varied Sittella, 32
Variegated Fairywrens, 177
Varricchio, David, 142–43
Vehrencamp, Sandy, 192, 193–96
Violet-backed Flycatchers, 51
Voelkl, Bernhard, 20–22
Von Schantz, Torbjörn, 153
vultures, 37–42, 247

W

Wallace, Alfred Russel, 47
warblers, 52, 53, 55, 71, 151–52, 153
Ward, Peter, 36, 39
Wattled Jacanas, 146
Weatherhead, Patrick, 28, 147
Weimerskirch, Henri, 20
West, Stuart, 181–82, 184, 185, 201
Western-crowned Warblers, 55
Westneat, David, 150
White-breasted Nuthatches, 46, 52, 58, 59
White-crowned Sparrows, 167
White-flanked Antwrens, 51, 61, 67
White-rumped Shama, 65
White-winged Choughs, 4, 210, 227–32, 244–45
Wideawake Tern, 104
Wiklund, Christer, 89
Wiley, R. Haven, 121
Willow Tits, 56, 57
Wilson, Alexander, 93
Wilson, E. O., 172
wing length, evolutionary changes to, 86–87
Winterbottom, John Miall, 51
Wion, Andreas, 244
woodcreepers, 62, 67
woodhoopoes, 189
woodpeckers, 46, 52, 53, 59, 60, 64, 94, 189, 210, 218–27, 235–36, 244–45, 248
Woolfenden, Glen, 170–71, 172
wrecks (die-offs), 112–13

Y

Yellow-billed Kites, 33
Yellow-headed Blackbirds, 72
Yellow-throated Minivets, 46
yuhinas, 5, 207

Z

Zahavi, Amotz, 36
Zebra Finches, 159
Zou, Fasheng, 64–65

About the Author

Joan E. Strassmann is an evolutionary biologist and behavioral ecologist known for her research on social evolution and cooperation. Her previous books include *Slow Birding: The Art and Science of Enjoying the Birds in Your Own Backyard* and *The Slow Birding Journal: A Field Diary for Watching Birds Wherever You Are*. She is the Charles Rebstock Professor of Biology at Washington University in St. Louis and is an elected member of the National Academy of Sciences and the American Academy of Arts and Sciences, is a Fellow of the Animal Behavior Society and the American Association for the Advancement of Sciences, and has held a Guggenheim Fellowship. She lives with her husband in St. Louis, Missouri, and Leland, Michigan.

About the Illustrator

Anthony Bartley is a Chicago-based visual artist and founder of FadingRoyalty and ~~HAPPY~~ Days Apparel. He has a passion for science, having been a member of several different types of research labs from age ten to undergrad. He is particularly interested in mental health in the Black community and the positive role that art can play. Find him at fadingroyalty.com.